CLIMATE, CHAOS AND COVID

How Mathematical Models
Describe the Universe

CLIMATE, CHAOS AND COVID

How Mathematical Models
Describe the Universe

Chris Budd

University of Bath, UK

World Scientific

NEW JERSEY · LONDON · SINGAPORE · BEIJING · SHANGHAI · HONG KONG · TAIPEI · CHENNAI · TOKYO

Published by

World Scientific Publishing Europe Ltd.

57 Shelton Street, Covent Garden, London WC2H 9HE

Head office: 5 Toh Tuck Link, Singapore 596224

USA office: 27 Warren Street, Suite 401-402, Hackensack, NJ 07601

Library of Congress Cataloging-in-Publication Data

Names: Budd, C. J. (Christopher J.), author.

Title: Climate, chaos and Covid : how mathematical models describe the universe /
Chris Budd, University of Bath, UK.

Description: New Jersey : World Scientific, [2023] | Includes bibliographical references and index.

Identifiers: LCCN 2022025783 | ISBN 9781800613041 (hardcover) |
ISBN 9781800613058 (ebook for institutions) | ISBN 9781800613065 (ebook for individuals)

Subjects: LCSH: Mathematical models--Social aspects.

Classification: LCC QA401 .B762 2023 | DDC 511/.8--dc23/eng20220829

LC record available at https://lccn.loc.gov/2022025783

British Library Cataloguing-in-Publication Data

A catalogue record for this book is available from the British Library.

For any available supplementary material, please visit
https://www.worldscientific.com/worldscibooks/10.1142/Q0385#t=suppl

Desk Editors: Jayanthi Muthuswamy/Adam Binnie/Shi Ying Koe

Typeset by Stallion Press
Email: enquiries@stallionpress.com

Preface

To a physicist, or an engineer, mathematics is a pale shadow of reality. To a pure mathematician, reality is a pale shadow of mathematics. But to a mathematical modeller, reality and mathematics are so intertwined as to be almost indistinguishable.

What is this book all about?

What is a mathematical model? Loosely expressed it is a mathematical simulation of the real world. Usually this is a set of equations, which when solved (often on a computer) tells us something about the real world that we didn't know before. Mathematical models have been much in the news recently. They lie at the heart of the methods that have been used both to predict the spread of COVID-19 and to then to help decide what to do to reduce its impact. Mathematical models are also used to predict climate change and guide us in helping to prevent it.

Mathematical models have been used for many years by engineers and scientists, and they lie at the heart of much of modern technology. For example, your mobile phone would not work without the mathematical formulae coded within it. We have also seen a recent, and significant, rise in the use of mathematical models to guide decisions made by our policymakers. Our weather forecast, and the predictions of climate change used to inform the Intergovernmental Panel on Climate Change, are made using a mathematical model. More recently mathematical models have been used in medicine, biology and the social sciences, and it is in this capacity that the conclusions from the models have been used (as we will see in Chapter 4) to advise policymakers engaged with the fight against COVID-19. However, this has led to widespread concern, and suspicion, about the use of the predictions mathematical models to make decisions.

It may surprise readers to learn that mathematical models, and indeed the idea of using mathematics to describe the real world, have been around for a very long time — certainly since the work of the Ancient Greeks, and probably before them. Isaac Newton described their use to explain the motion of the planets in his book *The Principia: Mathematical Principles of Natural Philosophy*, published in 1687, and they have been used heavily in physics and engineering ever since.

Mathematical models are carefully constructed by taking (known) laws of physics, biology, or social science, formulating these as equations, and informing this whole process by using observational data. The equations are then solved to give predictions. These predictions are then tested against reality, and the model is refined, improved, and extended, to the point where its predictions can be trusted to represent reality. However, regardless of how carefully it is constructed, a mathematical model is always an approximation of reality. Indeed it is remarkable that an abstract subject such as mathematics is as good as it is in representing reality at all.

As soon as you involve people in the modelling process (as we had to do in the modelling of the response to COVID-19) then things become much more complicated and uncertain. Nevertheless, using a mathematical model is much the best way to help predict what might happen in the future. The advantage of using a mathematical model over informed guesswork, is that you can be objective over the assumptions made in the model, including its powers and limitations. You can be quantitative in making predictions, which can then be tested. You can also run many 'what if?' scenarios to explore what might happen in different situations. Perhaps most importantly, although (as in all aspects of life) using a mathematical model involves a degree of uncertainty, you can quantify how much uncertainty there is in the model. Or to put it another way, you can say exactly how confident you are in its predictions. Modelling is about both solving the right problem, and also about solving the problem right.

We shall revisit these ideas repeatedly throughout this book, as we encounter many examples of constructing, testing, extending, and then using, a mathematical model. This approach will consistently follow a 9-Step plan which will be described in Chapter 1. The case studies will then be drawn from problems as varied as climate change, COVID-19, cooking a potato, going into space, and keeping the lights on. We will even use mathematical modelling to help us to cross the road and to save the whales. The title *Climate, Chaos and COVID* comes from three of the examples that we will look at.

Why did I write this book?

I have spent my whole career (including the gap year I spent before I went to university) using mathematical models to help understand the world. In the course of this I have worked in industry academia, and government. I have also had the honour of working (mathematically) with such great organisations as the Knowledge Transfer Network, the European Study Groups with Industry, the Smith Institute, the UK Met Office, BT, CCFRA, Airbus, Mondelez, the National Grid Company, the NHS, Pepsi-Co, Bristol, Amsterdam Zoos, the Royal Institution, Gresham College, Shakespeare's Globe and the Royal Opera House. My current position is 'Director of Knowledge Exchange for the Bath Institute for Mathematical Innovation'. The purpose of my work there is to lead a programme of research and teaching of mathematics as a means of exchanging knowledge between universities and industry.

During the course of my career I have worked on such varied problems as weather and climate forecasting, electricity supply, Wi-Fi optimisation, food manufacture, bee keeping, microwave cooking, and folk dancing. You will see examples of all of these (apart from the folk dancing) spread throughout this book. Since the UK lockdown in 2020 I have also been very busy working with the Virtual Forum for Knowledge Exchange in Mathematics (V-KEMS) which has been conducting a series of online workshops that focus the power of mathematical modelling to bear on the challenges presented by COVID-19. These have ranged from keeping our shops, workplaces, schools and universities safe from COVID-19, to helping the transport, healthcare, and leisure industries, recover from the devastating impact of COVID-19.

Doing this has given me a strong appreciation, and experience, of the power of mathematics to solve an incredible variety of real-world problems, and to do everyone a great deal of good in the process. This power has only been enhanced with the increasing use of computers both in society and also (since my first computing experience in 1971) in my own work. The potential future impact of mathematics is almost limitless. However, as I said at the beginning of this Preface, the use of mathematics in real world situations is often treated with suspicion. That is if it is considered at all. The use of mathematics in technology is often invisible. It is like the air we breathe. Without air we could not live, however we cannot see it, and often we forget that it is there.

I decided to write this book to try to explain some of the mysteries of mathematical modelling, and, hopefully, to dispel concerns of the use of mathematical models in the real world. I do this through a series of case studies, many of which are taken directly from my own experiences of working as a mathematical modeller in academia, in industry, and between the two. The book is an expansion of a series of 24 lectures that I delivered between 2016 and 2020 when I was the Gresham Professor of Geometry. This position, which is the oldest established chair in mathematics in the UK and has been running since the time of Elizabeth I, involves delivering six free public lectures a year on any mathematical topic. (Previous Gresham Professors have included Isaac Barrow and Robert Hooke, and more recently Sir Christopher Zeeman, Sir Roger Penrose, Ian Stewart, Robin Wilson, Raymond Flood, and John Barrow. The current professor is Sarah Hart.) I chose as my topic 'Mathematics and the Making of the Modern and Future World' exploring the way that maths has changed our world. The very positive response to these lectures led me to write not only this book on mathematical modelling, but also a further book on mathematical algorithms.

Who might read this book?

This books aims to show how mathematical models are constructed, what their power and limitations are, and what we can learn from them. It is richly illustrated by case studies taken from real world examples. The book examines the story of how mathematical models are used to model the real world, and the range of applications. In terms of the level of the material, this book is intended to be accessible to all with a general interest in the way that maths can be applied to the real world. It is not meant to be a textbook suitable for an undergraduate to learn all the details of modelling and associated mathematics (such as the theory of differential equations or of numerical analysis). Such topics are excellently covered in the texts of J. Murray [4], S. Strogatz [15], P. Drazin [16], D. Acheson [19], and (especially) S. Howison [47] that are listed in the References at the back of this book. These books tend in the main to look at simplified applications rather than examples which occurred in the context of real applications. Instead, this book is meant as a 'popular' introduction to the art and application of modelling which will try to make some sense of the way that the results of models are used to make decisions. The level of mathematics required will (usually) be that of a finishing high school student, if not less. If you like reading popular articles about maths (for example, in magazines such as *PLUS* or *Chalkdust*) or appreciate the videos on the Numberphile channel, then you should like this book, too. It is quite possible to read this book and skip the formulae in it, yet still gain an appreciation of the ways that models are used and the full context for their use. I do hope that you will then be enthused enough by the mathematics to read (at least) one of the textbooks referred to above.

Why are there equations in this book?

Any reader who picks up this book will find that it contains a fair number of equations. I hope by the fact that you are still reading this Preface, that you have not put the book down and walked away. I thought long and hard about whether to include equations in this book, and had many discussions with my publisher, about it. But this is a book about mathematical modelling, and the heart and soul of all modelling are the equations that we use to represent the world. You can think of an equation as a very concentrated piece of information. As the great Richard Feynman said in his 'Lectures on Physics' a formula, like Schrodinger's equation, contains within it a huge amount of information about the universe. The job of the mathematical modeller is not only to unlock this information, but to then use the solution of the equation to help to predict the future. So, I decided that I had to include the equations. However, adequate warning will be given in advance of each formula should you wish to avoid it, and guidance on where to start reading again without losing the sense of what is going on.

But I would be the first to admit that on first (and probably on second and third) sight, a mathematical formula can look rather scary and difficult to understand. I'm with you on this, I usually feel exactly the same way when I see one. But rest assured, help is at hand. The purpose of this book is to put the equations into the context of the real-life problems that they are there to solve, and to then unlock their secrets. We will be telling a mathematical story, with the mathematics providing us with the signposts to guide us on our way. I hope that you will enjoy the journey.

Another way to think about this journey is to imagine that an equation is like a strong glass of whisky (any classic malt will do). Sometimes you want to add it to a nice glass of spring water to bring out its flavour. I hope that this book will be like that glass of water.

Cheers!

Chris Budd OBE, FIMA, NTF,
Department of Mathematical Sciences,
University of Bath,
Bath BA2 7AY,
UK
December 2022

About the Author

Chris Budd OBE, PhD, FIMA, CMath, NTF, is currently a Professor of Applied Mathematics and the Director of Knowledge Exchange for the Institute of Mathematical Innovation at the University of Bath, UK. He did his first degree in mathematics at the University of Cambridge where he was Senior Wrangler (top first-class degree). He obtained his doctorate at the University of Oxford and later was a fellow of Hertford College. His first permanent position was at the University of Bristol, before he moved to Bath in 1995. He also holds the position of Professor of Mathematics at the Royal Institution and was the Gresham Professor of Geometry during 2016–2020.

His work is on the applications of mathematics to the real world, including climate change. In this capacity, he works a lot with industry, including the Met Office. Following the outbreak of the COVID-19 pandemic, he was one of the founders of V-KEMS (the virtual forum for knowledge exchange in the mathematical sciences), which played a major role in the use of mathematical modelling to fight the impact of the virus. In 2022 he received a SAMDS award for 'Modelling and Data Support' from the UK Government Office of Science, for his 'exceptional contribution to the work of SPI-M-O in modelling COVID-19'.

As a passionate populariser of mathematics, he gives many popular mathematics talks and workshops to schools, and other bodies, and for over 20 years has taught the methods of mathematics communication to undergraduates at Bath. In recognition of this work, he was awarded a National Teaching Fellowship in 2000, OBE in 2015, the Joint American Mathematical Societies award for mathematics communication in 2020 and the LMS 150th Anniversary Prize for the Communication of Mathematics.

When not doing mathematics, he divides his time between working with young people and going for long walks with his family and dog.

Contents

Preface   v

About the Author   xiii

1. How mathematics models real life   1

2. Why did the mathematician cross the road?   29

3. Mathematics saves the whales!   37

4. How mathematics helped in the fight against COVID-19   47

5. Can mathematics predict the future?   61

6. The mathematics of climate change   103

7. Energetic mathematics   145

8. Mathematical food for thought   175

9. Material mathematics   209

10. Mathematics goes into space   247

   References   275

   Index   281

1
How mathematics
models real life

1. Introduction

If were to ask anyone what are the great challenges for humanity in the 21st Century, it is very likely that curing major diseases such as COVID-19, and saving the environment, would be close to top of most people's lists. In this book we will address both of these problems, and many others, and will find solutions to them by using the powerful tool of *mathematical modelling*. Rather than being the useless subject, for which it is often portrayed, mathematics will be seen in this book to be able to make a real difference to many questions of major concern. Whilst some of the applications of mathematics are, without doubt, unpleasant, I would also hope that everyone would agree with me that saving the environment and curing diseases are without question positive things.

Whilst saving the environment and curing diseases are clearly very important issues in their own right, they are only two of the problems for which any possible solution involves the applications of mathematics. This book will explore many other problems where mathematics can help humanity. I hope that by the end you will be able both to use mathematics to solve, or at least give insights into, many major real life problems and will also understand the difficulties, uncertainties, and limitations, in using such a mathematical approach.

In this chapter I will explain the basic ideas behind the way mathematics is used to model reality, and will try to explain its power and its limitations. In the next chapters we will then look at how by using mathematical modelling we can use algebra to help us cross the road, geometry (including Pythagoras' Theorem) to help to save the whales, and differential equations to predict the progress of an epidemic such as COVID-19. Later chapters will explore the wide range of the ways that mathematical models are used in real life situations, from telling the future to predicting the climate, generating electricity, producing food, folding rocks, designing materials, and finally going into space.

2. What is a mathematical model?

Can maths describe the universe?

Mathematical modelling is the process of describing a real world problem in mathematical terms, usually in the form of equations, and then using these equations both to help understand the original problem, and also to discover new features about the problem which could not have been predicted without the use of mathematics.

> *A mathematical model is a mathematical analogue of a real situation, which is complex enough to be realistic and yet simple enough to analyse and make predictions with.*

Modelling both lies at the heart of much of our understanding of the world, and allows engineers to design the technology of the future. With modelling we can travel to the edge of the universe (both in space and time), peer into the heart of the atom, predict the behaviour of epidemics, understand the future of our climate, and the effects that climate change will have on our lives. Mathematical models can also be used to predict the behaviour of people in crowds, and traffic on a highway.

Hang on, you say, surely mathematics is an abstract creation of the human mind which on its own cannot solve any practical problem. The main requirement of mathematics is that it should be self consistent within the statements of its axioms, and it should lead, by rigorous arguments, to amazing theorems, such as Pythagoras' Theorem, which represent eternal truths.

To see why maths is so powerful in explaining the universe we need to ask ourselves the question: *Is the universe completely random, or does it have some order and pattern to it?* Answers on a postcard, please. In earlier generations the answer might have been that nature was essentially random, unpredictable, and subject to the whim of irrational deities. We now have a better answer.

Indeed it is very clear that nature does have underlying patterns behind it, and it was the realisation that this was the case, which started humanity on its modern course and led to the scientific revolution. It is precisely these patterns which are what scientists study. In fact I would strongly argue that

> *Science is defined to be the search for, and understanding of, the patterns that underlie the universe.*

So to understand the universe we need to study its patterns.

> *It is mathematics, which not only underlies these patterns themselves, but also gives us a way of describing them, and predicting them in advance.*

It was realising this that changed the way that we see the universe and has led to the modern world.

If we open our eyes we can see mathematical patterns all around us. A great example is the snowflake, of which eight are illustrated in Fig. 1.1. It is immediately clear from this picture that while any two snowflakes are different in many ways, each of them has a beautiful, marked, and precise *six-fold geometrical symmetry.*

We are so used to this beautiful symmetry that we may never stop to wonder at it. But a good scientist/mathematician or indeed any of us *should never lose our sense of wonder!* What is going on here is truly remarkable. The snowflakes have clearly paid attention to their geometry classes at school, and it is very clear that they are obeying mathematical laws. In reality a snowflake is formed by water droplets freezing high up in the atmosphere. These droplets are in turn made up of tiny molecules which are (relatively) huge distances from each other. However, the molecules in one part of the snowflake seem to know exactly what those one sixth of a turn away are doing, and behave in the same way! Here we see a high degree of pattern and regularity. The reason being that the laws of thermodynamics have a sixfold symmetry built into them, and these

Figure 1.1 *All snowflakes are different, but they all have a six-fold symmetry. (Credit: https://www. shutterstock.com/image-photo/collection-snowflakes-happy-new-year-natural-114893878)*

laws acting on the molecules (which themselves have a high degree of symmetry) lead to snowflake shapes with low energy which inherit this symmetry and pattern. We can *predict* that a snowflake will have this symmetry, even if we don't know its exact shape. In a very similar way we will find that we can predict the general form of the climate in 10 years time, even if we can't predict the weather on any day of that year.

A swinging example of mathematical modelling

One of the first of the true modern scientists to realise that nature followed mathematical laws was Galileo Galilei of Pisa (1564–1642) (see Fig. 1.2).

Galileo is one of my favourite scientists for a number of reasons. We share the same birthday (but not the same year!), and I admire him both for his scientific insights, and his incredible courage in pursuing these in the face of extreme opposition. One of his more famous quotes is that *"the Universe is written in the language of mathematics"*. I could not agree with him more. Galileo made

Figure 1.2 *Galileo Galilei and the chandelier in Pisa Cathedral that led him to discover the regular motion of the pendulum. (Credits:Left: https://commons.wikimedia.org/wiki/File:Galileo.arp.300pix. jpg. Right: Vitold Muratov, https://commons.wikimedia.org/wiki/File:Pisa.Interier_of_Cathedral.jpg)*

significant discoveries in kinematics, and in astronomy (using the newly invented telescope), as well as studying the strength of materials. Many of his discoveries were made in Padua where he taught geometry, mechanics and astronomy from 1592. According to Isaac Asimov [1], "It was after accidentally attending a lecture on geometry, he talked his reluctant father into letting him study mathematics and natural philosophy instead of medicine". (One wonders what other effects lectures on geometry might have!) However, it was before then in 1581, when Galileo was still studying medicine, that he noticed (during a dull sermon) a swinging chandelier in the cathedral in Pisa (see Fig. 1.2). What impressed him about this was that as it swung in larger and smaller arcs due to the effects of air currents the chandelier *took the same amount of time to swing back and forth*, no matter how far it was swinging. One of the chandeliers at Pisa Cathedral is pictured above next to Galileo. (For other things to do during a dull sermon see the excellent book by T. Sims *et al.* [2].)

Galileo timed the length of the swings of the pendulum with his pulse, and then checked his observations later with careful experiments at home. In his experiments he looked at pendulums of equal length but with different sizes of swings. By doing this Galileo showed that the swing time of a pendulum did not depend upon where it was, or on its position, or even the time of the day. All that mattered was its length. Thus the swing of the pendulum was predictable. This was an astonishing observation. One hundred years later this observation led Christiaan Huygens to develop an accurate pendulum clock.

What Galileo did not know at the time was why this was true. To find out we had to wait (not for very long) for another great scientist. Isaac Newton (1643–1727) was born just after Galileo died, and partly building on Galileo's work, went on to transform our understanding of the universe. He did this both through his discovery of many of the laws which lay behind the patterns of the universe, and just as significantly, for his invention (along with various others) of the mathematical technique of calculus, which gives us the essential tool to understand and exploit these laws. Chief amongst these discoveries were those of the *three laws of motion,* which gave a clear description of how a moving body would behave. These laws could in turn be expressed in mathematical terms, particularly as *differential equations* which gave a precise mathematical description of how they would evolve in time. We will meet differential equations in the SIR models for COVID-19 in Chapter 4, in the models for climate change in Chapter 6, and many times later in this book.

We will use the example of the pendulum given by the swinging chandelier described above to illustrate the process, and power, of *mathematical modelling.*

To do this we must start by considering the physical process we are trying to study and then try to turn this into formulae which we can analyse using mathematics, and which can then tell us something new.

We start by thinking about what a pendulum is. For a simple model we will treat the chandelier pendulum to be a heavy weight of mass m (the chandelier) on the end of a taught wire, which has a length l. The chandelier then swings about a pivot at an angle θ to the vertical. The pendulum and the forces acting on it in various directions are illustrated in Fig. 1.3.

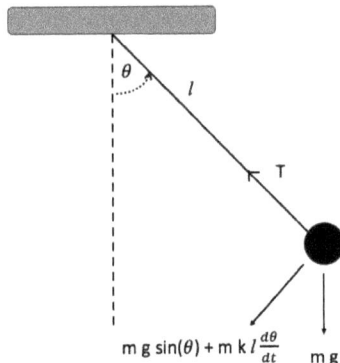

Figure 1.3 *The forces acting on a swinging pendulum.*

Now we consider the physics of the problem. In its swings it is acted on by a constant downwards gravitational force of mg (where g is the acceleration due to gravity). The other forces acting on it are the tension T in the wire, together with a small retarding force acting on it proportional to the *speed* of the chandelier (where we will take mk to be the constant of proportionality), which combines the effects of friction and air resistance and is always in the direction opposite to that of the moving pendulum. The forces are also shown in Fig. 1.3.

To make progress we now have to turn this picture into mathematical formulae.

At this point I have to issue a warning

> *Danger Mathematical Equations at Work.*

We are about to encounter some fairly tough mathematical equations. They are tough partly because the best way to describe physical phenomena is usually through the medium of differential equations. These are equations which describe how things change in time and space. Differential equations have been around for about 500 years and they are a fabulous way of describing the real world. However, even simple differential equations can be really hard (if not impossible) to solve, even for professional mathematicians. So if you find them heavy going, then you are in excellent company. A quote often made about any attempt to put mathematics into a book (such as this one) is that every equation in it halves the sales. I hope that this is not the case for this book as there will be a lot of (differential) equations in it. But please bear with me, they really are the best tool that we have for understanding the universe, and we would be surprised if nature did not give up its secrets easily.

To find the differential equations of motion for the pendulum we will look at the motion of the chandelier in a direction at 90 degrees to the wire, as the tension force T has no component in this direction, in contrast the retarding force due to air resistance is all in this direction. and this retarding force is given by mk times the speed of the pendulum. The speed of the pendulum itself is given by $l\, d\theta/dt$. Hence, the retarding force due to air resistance is given by the expression $mkl\, d\theta/dt$. The component of the gravitational force in this direction is given by $mg\sin(\theta)$. Both of these forces act in the direction opposite to the motion of the chandelier. The complete force acting in the direction of increasing motion of the chandelier is shown in Fig 1.3 and then given by

$$F = -mkl\, d\theta/dt - mg\sin(\theta).$$

Here is our very first differential equation, For those of you unfamiliar with differential equations (which I assume will be most of you) the term $d\theta/dt$ in this expression is called the *first derivative* of θ with respect to time. It is an expression used in the branch of mathematics called calculus to describe how rapidly the angle θ is changing with time. We will be meeting derivatives frequently throughout this book. Calculus forms the bedrock of mathematical modelling.

Now Newton's Second Law relates forces to accelerations through the equation

$$F = m\,a,$$

where a is the acceleration of the chandelier in the direction of motion. In differential equation terms the acceleration is expressed in terms of a *second derivate* of θ with respect to time given by the expression.

$$a = l\frac{d^2\theta}{dt^2}.$$

Combining everything together, and dividing through by l and by m, we arrive at an equation in terms of the angle θ for the motion of the chandelier

$$\frac{d^2\theta}{dt^2} + k\frac{d\theta}{dt} + \frac{g}{l}\sin(\theta) = 0.$$

Bingo. We have created our first mathematical model in this book. As advertised earlier, it is a description of the motion in terms of a *differential equation* which tells us how the *angle θ* changes with respect to the *time t*. The solution of this equation should mimic the real-life behaviour of the pendulum. All we need to do now is to solve this equation to predict the future. As I said earlier, differential equations lie at the heart of nearly all of mathematical modelling. They are the tools which explain how the universe changes. Many mathematicians (including myself) have spent their whole lives studying them. They are both the key to understanding the universe, and are also (as we might expect) frustratingly hard, if not impossible, to solve.

So ... as the doctor says, I have some good news and some bad news for our particular differential equation.

The *good news* is that the differential equation above works. It really does describe the motion of the pendulum *for all time*. And it does this very well. Using this equation we can predict the future motion of the pendulum unless something else (such as a human being) changes its path. Amazing.

Now for the *bad news*. The equation above is an example of a *nonlinear ordinary differential equation*. These equations are phenomenally hard to solve exactly at the best of times (I have spent most of my career trying, usually unsuccessfully, to solve them). Certainly this equation is well beyond most university undergraduate courses. (To solve it exactly we need to use Elliptic integrals and other heavy duty mathematical machinery, and even then there is a lot that we still cannot solve.) Nature is like this. When we write down mathematical models we often end up with equations that we can't solve using the mathematics that we currently know. Let no one tell you that applying mathematics to the real world is easy, or that it is simply a matter of taking known mathematics and applying it to your problem to get a result. Solving real world problems is (usually) hard, and requires a lot of creative thinking. In the long term this is a *good thing*. Working out how to solve hard problems is one of the great drivers of much of the development of mathematics, and the maths that we develop to solve them usually has many other applications. Also don't worry if you get stuck with trying to solve a piece of mathematics. This is normal with a real world problem, and the normal state of any mathematician is to be stuck. However, if we cannot wait long enough for an exact solution to be found by some great mathematical breakthrough, then we need to find practical ways of solving the equation. One excellent way is to use a computer to solve it and a lot of effort (and excellent mathematics) has gone into creating computer algorithms which can solve differential equations. Figure 1.4 is an example of two computer solutions of the motion of the pendulum, the top one for a small amplitude swing, and the bottom for a much larger swing. I am sure you will agree both do look like the motion of a swinging pendulum which eventually comes to rest.

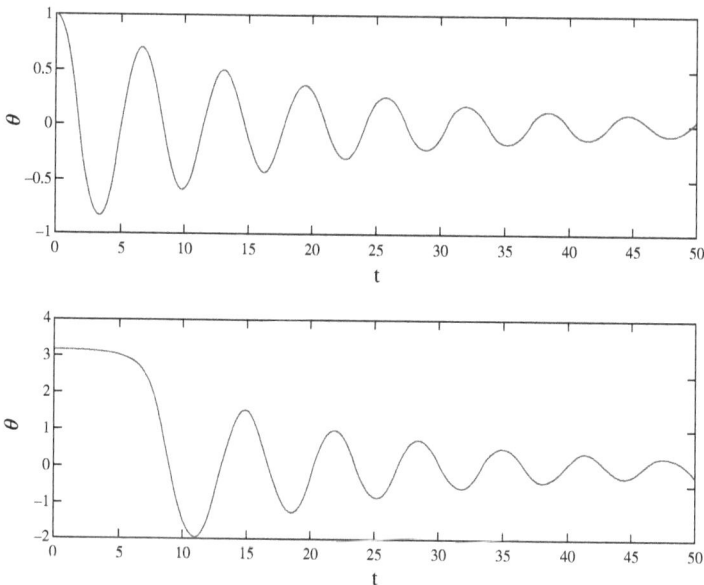

Figure 1.4   *Two computer simulations of the damped pendulum showing how the angle changes with time. The top figure shows fairly small swings, and the bottom figure shows larger swings.*

I am a great fan (and a proponent) of this way of doing things. Computers are a vital tool to help us solve hard mathematical problems. But even I have to admit that using a computer not only brings in errors but also means that it is hard to gain insight into the general behaviour.

Another approach, which lies at the heart of mathematical modelling, is to look for ways in which we can *simplify* the problem to something that we can solve, whilst still retaining its essential physics, so that the simplification still gives us useful insights into its behaviour.

One way to do this is think of what was actually happening to the pendulum that Galileo was actually looking at. His pendulum was being slightly disturbed so that it has *small swings*. Also it was heavy, and as a result the air resistance on it was small. The small air resistance means that we can approximate $k = 0$, and the small angle of swing means that we can use the approximation

$$\sin(\theta) \approx \theta.$$

The differential equation for the pendulum then simplifies to:

$$\frac{d^2\theta}{dt^2} + \frac{g}{l}\theta = 0.$$

Now for more good news. If the pendulum starts from rest then this equation has an exact solution (discovered by Euler). This is given by:

$$\theta = A\cos\left(\sqrt{\frac{g}{l}}\,t\right).$$

Note the remarkable nature of this solution. It uses the cosine function which is normally associated with trigonometry and triangles. But here we see it as a way of explaining a physical phenomenon which has nothing to do with triangles. I find this truly remarkable and a wonderful illustration of the power of mathematics. This simple formula predicts a lot about the motion of the pendulum.

Firstly that the pendulum swings to and fro with an amplitude A, just as we would want it to.

Secondly that if A is small enough for this approximation to be valid, the period of the swing is given by:

$$T = 2\pi \sqrt{\frac{l}{g}}.$$

This formula predicts that the period of the (small) swing is *independent of the size of the swing*. Amazing! This prediction agrees completely with Galileo's observations in Pisa Cathedral.

It also tells us how the period changes with the length of the pendulum. So even if we have never built a pendulum with a length of 100 m (or indeed 1 km) we can still say to high precision how long it will take to swing.

As an amusing note. The acceleration due to gravity is given in SI units as $g = 9.81$ metres per second per second. If we take a one metre long pendulum so that $l = 1$ m, then $T = 2.00607$ s, so that the time for half a swing (i.e., from end to end once) is *almost exactly one second*. I have often wondered whether it is mere coincidence that in SI units the square root of g is almost exactly given by π? It seems too good to be true, but I can think of no reason why it should be so.

The success of this mathematical approach gave Galileo and Newton (and many others including Leonardo da Vinci) an idea for a general approach to understanding the Universe which goes something like the following:

- Write down the mathematical equations describing a physical system
- Solve the equations
- Using your solution, predict the future

This was a truly ground breaking idea. Its invention was perhaps the greatest moment in science ever, and of course is the motivation behind this book!

Mathematical modelling in a bit more detail

The techniques that we used to understand the behaviour of the pendulum are just one example of mathematical modelling and is the job that I have been doing for my entire career. Mathematical models can be used both to help understand a situation, and also to predict (in some cases) what the system that you are modelling will do next. Such mathematical models were heavily used both in the COVID-19 epidemic and in decisions on climate change, to advise governments and many other people who make policy. There will be examples of them in much more detail later in the book. These applications of mathematics bring mathematicians into situations involving ethical, moral and legal judgements, which directly affect people's lives and which seem a long way away from the cold, hard, formulae in mathematical textbooks.

The techniques that we use in a modern mathematical model are an extension of those used by Galileo and Newton and can be summarised by the following nine steps:

Step 1	Think about the **key principles** involved in the system that you are trying to describe, using as much data as you have to give you insight.
Step 2	Think about the **basic physical or social laws** governing these principles.
Step 3	**Simplify** as much as possible. But no more!
Step 4	Formulate the principles as *mathematical formulae*.
Step 5	*Solve* the mathematics to **make predictions**, *computing* where necessary.
Step 6	Constantly *compare* the model with data and calibrate it where necessary.
Step 7	*Update* the model to make sure that it agrees with the data. Repeat from Steps 1, 2 or 3, adding, improving, or reject parts of the model till things look correct.

Step 8	**Communicate** the predictions in a way that they can be understood by their intended audience, making the audience also aware of the uncertainties and limitations of the model.
Step 9	Using your shiny (validated) model, *make many more predictions*, design things, predict the future, advise governments etc., always being aware of the uncertainty inherent in any models and of the (ethical) implications of the results of using the model to make decisions.

That all looks simple enough, and we will follow these nine steps closely when we develop the models for various problems later on in this chapter, and then later on in this book. But in practice mathematical modelling is almost as much an art as a science. One reason for this is that it is rare to come anywhere close to writing down the right equations the first time. Indeed without looking hard at the data to start with, it is likely that the equations will not be anywhere close to the truth. The result is sometimes the construction of 'mathematical models' that whilst they look nice are often so far from the truth as to be practically useless. They are also often so simplified, that they have no real mathematical interest either. In contrast the true process of mathematical modelling plays close attention to the data at all stages of the process, employs computation at all times (on often very hard formulae), is (acutely) aware of the limitations and uncertainties in the model, and NEVER stops at Step 7 above. A mathematical model is a living process, that if looked after well will continue to give insights into the system. Another problem with this approach is that Step 6 is often very difficult. What does 'agreeing with data' really mean when it comes to a model of (say) loneliness. The best models are ones which give us excellent insight into the system which allow us to make useful future predictions. Quantitative agreement with actual data is often a bonus. Or to quote George Box [3]

All models are wrong, but some of them are useful.

The process of simplification in Step 3 of constructing a model is (as we have seen) also hugely important, but also hard. We will be exploring it a lot in this book. A wonderful quote from Albert Einstein is:

> *A model should be as simple as possible, and no simpler.*

Einstein shows strong insight here, but I'm sure that he was well aware that knowing when a model is just simple enough is very hard. In practice one way to simplify is to make estimates of the importance and size of the different processes affecting your system, and to then ignore those which are too small to make much of a difference. For example (at a physical level) we do not compute sound waves when we are making a weather forecast, and in the case of the pendulum (at a mathematical level) we ignored the small difference between $\sin(\theta)$ and θ when the amplitude of the swings was small.

Finding the right simplification is almost the most vital part of the art of mathematical modelling. Finding a small system of equations, which capture the essential essence of it, and, crucially, are simple enough allows us to make analytical calculations. A formula derived from an analytical calculation can then give a clear view of the role of the parameters in that system without having to run a very large number of calculations. As we saw with the pendulum one single formula allows us to work out the period for any length of pendulum. Indeed, all good mathematical models have this ability to predict how a system will behave for a wide range of its parameters.

The part of mathematical modelling which most mathematicians will recognise (and is in their comfort zone) are Steps 4 and 5 above where we write down the mathematics and then solve it. Almost any area of mathematics can end up in a mathematical model! In the example above we have used differential equations. These (in both of their flavours as ordinary and partial) form the bed rock of much of mathematical modelling. However, as we shall see in this book, geometry, trigonometry, arithmetic, (vector) calculus, analysis, and algebra, all play vital roles. Also, increasingly in a world where we have to deal with data and uncertainty, statistics and probability are vital parts of most modern mathematical models of reality. It is the way that not only do these areas of maths find a role in mathematical models of reality, but also that modelling reality leads to new mathematics, makes the subject so interesting, powerful and important.

Important Note What I have described here is a mathematical model based upon the physical principles identified in Steps 1 and 2. As such it is a *physics driven model*. All of the models in the rest of this book will be like this. However, a lot of modern simulations of reality are made by *data driven models*. In these a computer algorithm (such as a neural net) is trained on a set of data to recreate reality. This method is at the heart of machine learning. Machine learning is growing rapidly and there are many books about it such as Wilmott's very accessible account in [7].

3. Some success stories of mathematical modelling

Newton's law of gravity

Perhaps the earliest example of a mathematical model of enormous predictive powers was Newton's law of gravitation applied to the Solar System. Rather than simulating the whole system in all of its complexity, he introduced the huge (but very effective simplification) of treating the Sun and the planets as single points. This allowed him to write down the basic equations of motion of the whole of the Solar System. His model became even simpler when he considered the Sun as being very heavy and not being affected by the other planets, and that each planet was in turn only affected by the Sun and not by the other planets. This allowed him to look at a set of 'two-body problems' for each planet, which simply looked at the model of a 'point' planet orbiting a 'point, but very heavy' Sun.

Solving these simple two body problems gave enormous insights! These included working out why planets went round the Sun in elliptical orbits, why Kepler's Second Law that equal areas were swept out in equal times was true, and perhaps most tellingly, the simple relationship that the cube of the orbital distance R of a planet from the Sun, was proportional to the square of the period T of its motion around the Sun. This is called Kepler's Third Law. If T is measured in years and R in astronomical units (AU) (the distance of the Earth from the Sun), then $R^3 = T^2$. Figure 1.5 demonstrates Kepler's Third Law in this case.

Here we can see the difference between a model and a simulation. If we want to simulate the motion of the planets, and other bodies in the Solar System, then we can write a big computer programme to do this very accurately. If we wanted to find out the period of a new planet then we would have to run this again, and then again and again for more and more planets. However, the formula above means that we can instantaneously predict the period of any new planet without having to run a new simulation. It therefore saves a vast amount of computing effort.

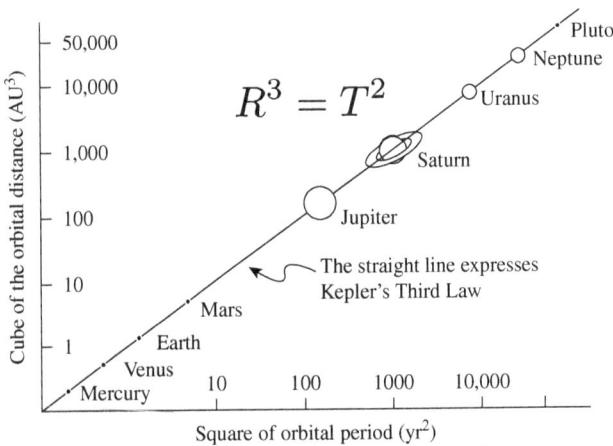

Figure 1.5 *A log-log plot of the cube of the orbital distance (in AU) of the planets as a function of the square of the orbital period in years. This graph is a straight line, which demonstrates Kepler's Third Law.*

The discovery of Neptune

After Newton had written down the laws of motion for the Solar System they were found to be very accurate in predicting the paths of the then known planets. However, on 13 March 1781 (the date, though not the year, of my wife's birthday) the planet Uranus was discovered by William Herschel from his garden in my hometown of Bath, England. When its orbit was plotted it was found that whilst it nearly agreed with the predictions of Newtonian mechanics, there were small discrepancies. At this stage the trust in the accuracy of the underlying model was such that even these small discrepancies caused a large amount of concern. It was postulated that there must be a cause, and one explanation was that there was another planet, which was disturbing the orbit of Uranus. Using the Newtonian model for the Solar System, but making it more complex by adding the effect of one planet on another, it was possible to calculate the location of this planet. This calculation was carried out independently, by John Couch Adams at Cambridge, and by Urbain Le Verrier in Paris. Both obtained very similar answers, and in response to Le Verrier's calculations Johan Galle using the telescope in the Berlin Observatory discovered the planet in 1846. We now call this planet Neptune, and it is illustrated in Fig. 1.6 with Uranus on the left.

As a sequel to this story, when better data came along about the orbit of Mercury, it was found that there was a very small difference between the orbit predicted

Figure 1.6   *The planets Uranus (left) discovered by William Herschel using a telescope in Bath, and Neptune (right) the existence of which was predicted by using mathematics and discovered later.*

by Newton's laws and the observations. Astronomers of the time postulated the existence of another planet, Vulcan, close to Mercury, to explain this discrepancy. But no such planet was found (sorry, Mr Spock). Thus Newton's model failed Step 7 and something had to be done to fix things. The result of this was the development of Einstein's General Theory of Relativity in 1915, which completely explained Mercury's orbit. Einstein's theory extends Newton's and predicts new phenomena such as Black Holes. We will look at how maths goes into space in more detail in Chapter 10.

By looking at the patterns of the Universe, converting them to mathematics and solving the equations, it was possible to predict the existence of something which had never been seen before! This is quite remarkable. This great triumph gave mathematicians confidence that they could predict many other phenomena, and to a large extent they were right. An early example of this was using mathematical formulae to predict the tides to high precision (using an early computer). Later in 1860 James Clerk Maxwell, by formulating Faraday's laws of electricity and magnetism as mathematical equations, and then solving them, was able to predict the existence of electromagnetic radiation many years before it was found experimentally by Heinrich Hertz.

Weather forecasting

The modern methods of weather prediction work by taking today's weather and then seeing how it evolves by solving the *Navier–Stokes equations* for atmospheric motion. These are given by:

$$u_t + u.\nabla u = -\nabla P + \frac{1}{Re}\nabla^2 u, \quad \nabla.u = 0.$$

Again, do not be afraid. I know that this is a very scary-looking equation. It is also very hard to solve. But this equation really does describe the weather very well. In the equation, u is the velocity of the fluid (the wind) and P is the pressure of the air. This equation is a lot more complicated than the earlier differential equations. The upside-down triangles in the equation are a mathematical shorthand to express the way that u and P change from one point in *space* to another. When we combine them with the term u_t which is a shorthand for how u changes in time, then we get an example of a *partial differential equation* which describes how u and P evolve in both space and time. Partial differential equations will star in later chapters of this book as well.

The complexity of these equations means that they have to be solved by a (super) computer. This can now be done accurately enough to predict tomorrow's weather (such as that illustrated in Fig. 1.7) with high accuracy.

Despite rumours to the contrary, this process works and works well — at least for short-term weather forecasting. We will look at it in much more detail in Chapter 6 when we examine climate modelling.

Strictly speaking a weather forecast is a *simulation* rather than a model. The difference between a simulation and a model is that in a simulation we try to get all the details as right as possible so that the conclusions are as accurate as possible and no simplifications are used. Using such simulations we can, for example, determine in advance whether a bridge will stay up after it has been built. We can also test the bridge to destruction without ever having to build it in the first place simply by varying the parameters in the computer

Figure 1.7 *A satellite image of the weather over the world. This can be predicted accurately by using a weather forecast based on a mathematical model.*

simulation. Simulation is used in the training of pilots in aircraft simulators, which are designed to be as close to reality as possible. Using these a pilot can be trained to fly an aircraft and to deal with dangerous situations, long before they have to enter the cockpit. Whilst simulators are very useful they have big disadvantages. The need for high accuracy means that the equations are usually far too hard to solve analytically. Instead, they must (often) be solved by using large supercomputers. The bigger the computer the better. These simulations often take a long time, consume a lot of energy, and produce vast amounts of data. So much data in fact that it is often hard to work out what is important and what is irrelevant. Furthermore, it is hard to use the simulations to do 'what if' experiments as they take so long to run and are expensive, and this makes it harder to quantify the uncertainty in the model. A second disadvantage is that they tend to only work, and be applied to, problems where the basic science is well understood. One reason for this is that it is a very significant amount of work (in person hours) to write and code up a simulator. You do not want to put such an investment into a system which you do not understand well.

Despite these disadvantages, simulations currently have a big advantage over the data driven models that are often used in machine learning applications. By being based on well understood physical principles, simulators have a level of trustworthiness and explainability, which is not present in a machine learning algorithm. They are thus inherently more reliable, and better able to cope with unusual circumstances. It remains to be seen how long this advantage will continue.

Biological sciences

It has taken longer for mathematical modelling to make an impact in the biological sciences. This is mainly because biological systems are inherently much more complex than physical ones (consider modelling a cell or the brain for example). It is also generally harder to make, repeat, and quantitatively assess, experiments on biological systems. Nevertheless, by a lot of careful study, significant progress has been made in recent years,

starting with pioneering work by Alan Turing just after World War II. Mathematical models are now used to help understand changing animal populations, the evolution of biological patterns, the spread of disease and the functioning of the nervous system. Indeed there is now a (relatively) new subject called Mathematical Biology, and a great overview of this is given in [4]. A major success of mathematical modelling applied to the biological sciences is in the field of epidemiology. I will look at a significant example of the use of mathematical biology in the study of the spread of COVID-19 in Chapter 4.

Social sciences

Modelling is both useful and important in the social sciences but is complicated by the fact that we have to take human behaviour into account. Given the complexities of the way that humans think and behave, this is generally difficult, if not impossible. Personally I am very suspicious about attempts to use mathematics to model aspects of human behaviour such as love and relationships. Although, have a read of *The Mathematics of Love* by H. Fry [5] and decide for yourself.

During the COVID-19 pandemic it was essential to combined the epidemiological models described above, with models of how people would respond, both to the pandemic and also to the government regulations aimed at controlling it. In the UK one of the bodies responsible for doing this modelling was SPI-M (Scientific Pandemic Influenza Group on Modelling). They faced a challenging job. In a published article [8], the chair of SPI-M, Professor Graham Medley said:

> *Forecasting weather is easier than forecasting disease, because the weather doesn't rely on what people do. It happens anyway. We aren't relying on the fact that the amount of rain depends on how many people take umbrellas with them.*

Wise words indeed.

Various areas in the social sciences where mathematical modelling has also made an impact are in economics, sociology, and in the behaviours of large groups. Whilst the behaviour of an individual is very hard to model, the collective behaviour of a large group, such as a crowd, or a city, becomes much more predictable. This is because the random actions of the individuals average out over the large group. Furthermore the behaviour of an individual in a crowd follows certain rules, and the collective dynamics of the crowd can be predicted as a result. Such calculations are very useful when it comes to designing sports stadia or railway stations. In the wonderful *Foundation Series*, Isaac Asimov wrote about a future in which, using the mathematics of Psychohistory invented by Hari Seldon, the behaviour of entire civilisations could be predicted by using mathematics. We are still a long way from this being possible.

The rest

In my career I have been asked to model many things, including the reaction of homeless people to changes in government policy, microwave cooking, fish in an aquarium, organ playing, cocoa growing, and the transportation of biscuits in a factory. I take all of these seriously and do my best. In fact we will meet some of these models in later chapters of this book. However, I draw the line at questions (usually posed by journalists when there is not much news around) such as what is the ideal shape for a Christmas tree, what is the perfect kiss, and what is the perfect joke? To be honest, I don't really think that mathematical modelling can really help answer such questions.

4. Common mistakes in mathematical modelling

I hope that you now appreciate how we can construct a mathematical model, and why this is a useful thing to do. But before we end this chapter, and also before we look at more examples of mathematical models in the following chapters, we will consider some of the common mistakes and limitations in the use of mathematical models.

Don't eat the menu

All models, even complex simulations, are approximations of reality. They are not reality itself. This should always be borne in mind when making predictions.

It's far better to say "that my model for climate change predicts a rise of three degrees in temperature" than to say that there will be a three degree rise.

> *When using a model always be aware of the assumptions, limitations, uncertainties and approximations that go into them. The best models have a quantification of their uncertainty. Constantly question these assumptions, and be prepared to change them if the model predictions deviate significantly from reality.*

Having said that, for all the benefits and limitations of mathematical or statistical models outlined, not to employ them is at best educated (and at worst uneducated) guesswork.

The mathematical drunkard

It is a very common mistake in mathematics (and indeed in most things) to change a problem to something that we can solve, rather than solving the original problem because it is too hard. In doing this we are behaving rather like a drunkard who drops a coin in the street. Rather than looking for it where they dropped it, they look for it under a street lamp some distance away. When asked by a police officer why they are looking there, rather than where they dropped it, they say that the light is much better under the lamp.

A side effect of this is that we might end up using overly sophisticated maths (or indeed the wrong sort of maths altogether) to solve what is in effect the wrong problem.

> *Mathematical modelling is all about firstly solving the right problem, and only then to solve the problem right.*

The ivory tower

Another common mistake (linked to the one above) is to have such a high opinion of what we know that we think that our knowledge on its own will solve any problem. To quote J. L. Synge in the *American Mathematical Monthly*, Vol. 51 (1944):

> *Nature will throw out mighty problems, but they will never reach the mathematician.*
>
> *He may sit in his ivory tower waiting for the enemy with an arsenal of guns, but the enemy will never come to him. Nature does not offer her problems ready formulated. They must be dug up with pick and shovel, and he who will not soil his hand will never see them.*

Many of my mathematical colleagues have asked me to find them industrial companies which can simply make use of their skills, or believe that all the maths needed to solve a problem has already been created and just needs to find the right application. It doesn't work like that. Although there are rare exceptions, mathematical modelling is a product of teamwork, with close collaboration between the mathematician, many other scientists, and the end user who needs the results. Only by having a certain amount of humility and a willingness to 'get stuck in' will a mathematician make any progress in modelling the real world.

There are fortunately great mechanisms in place for bringing mathematicians and users of mathematics (such as industry) together. These are called Study Groups or Think Tanks. See M. Barons' article [6] for details on how they work and how to get involved. The set up and operation of study groups all around the world as a way of bringing mathematicians and industry together have been one of the true success stories of the last fifty years, and the use of virtual on-line study groups to solve COVID-19 related problems played a valuable role during the pandemic.

The curse of the formula

The examples of models that I have shown above have led to what mathematicians might consider to be some fairly simple formulae. But don't assume that this will always be the case, and certainly don't over simplify the problem to get to an exact formula. Be happy, and prepared to use approximation techniques if they yield useful answers. These include asymptotic and numerical methods.

5. What next?

Given that so much of the real world can be modelled using mathematics, the potential applications of mathematical modelling are almost limitless. We will now explore this wonderful realm of possibilities. In the next three chapters we will look at three very different case studies in which we use a range of mathematical ideas to model real life. In these we will go through the process of building a mathematical model, and then using it to ask some 'what if?' questions. The first example uses a bit of algebra to cross the road, the second some geometry to save the whales, and the third some differential equations to track the course of the COVID-19 pandemic. In the later chapters we will then go on to explore some substantial applications of mathematical modelling in depth.

But I will finish this chapter with a singularly appropriate quote from the late and great Douglas Adams which not only links to the case study that we will look at in Chapter 3, but puts the whole of our endeavours to understand the universe into perspective:

> "On the planet Earth, man had always assumed that he was more intelligent than dolphins because he had achieved so much — the wheel, New York, wars and so on — whilst all the dolphins had ever done was muck about in the water having a good time. But conversely, the dolphins had always believed that they were far more intelligent than man — for precisely the same reasons".
>
> Douglas Adams, The Hitchhiker's Guide to the Galaxy

2

Why did the mathematician cross the road?

1. Overview

If asked 'Why did the mathematician cross the road?' you might come up with several answers from 'to see it from another angle' to 'for some complex reason' or maybe 'to prove that the other side exists' (one for the analysts amongst you). In this chapter we will answer the question by saying that 'they did it to stretch their modelling legs'.

But more seriously, what is the best way to cross the road, and can mathematics find it? We will look at a simple model to do this which only involves a bit of *algebra* (and some reasonable assumptions on the way that people will behave when crossing the road). And we first have to decide what we mean by best.

Of course we are all highly careful and law abiding citizens, and therefore we only cross the road at a pedestrian crossing. Most of the crossings where I live in Bath are those where you press a button and wait for the traffic light to turn red and the pedestrian light (typically a 'flashing man') to turn green, accompanied by a lot of beeping. Similar crossings exist all over the world.

I'm sure that you will all have noticed that when you press the button there is a delay of length D before the light turns green. The reason for this delay is to make sure that the traffic isn't blocked by a constant flow of pedestrians. But the question is: how long should the delay be?

Too short and the traffic is blocked. Too long and the pedestrians give up waiting and chance it by running across the road. There must be an *optimal time*. But the question is, what is it?

2. Making a basic model of crossing the road

To answer this question we can try using a simple mathematical model for the crossing process, using the steps to creating a model that we described earlier.

Figure 2.1 A typical pedestrian crossing in the UK controlled by traffic lights.

Step 1 We start by thinking about the basis issues that we encounter when we cross the road using a pedestrian crossing. A picture always helps to get our thoughts together (see Fig. 2.1).

After a certain amount of thinking we might come up with the following simple set of initial assumptions for what happens:

1. Pedestrians arrive at the crossing at a steady rate.
2. Traffic arrives at the crossing at a steady rate.

3. It takes the pedestrians a time T to cross the road.
4. The ideal delay time is the one that maximises the average total flow F of both the pedestrians and the traffic.
5. If the delay is not too long then the pedestrians will wait till the 'green man' starts to flash. However, if the delay is too long then they will take their lives into their hands and will start to cross anyway. Doing this they will endanger their lives, and will disrupt the traffic flow, which has to slow down to avoid them.

Step 2 Next we think about some of the physics behind these assumptions.

By a *steady rate*, we will assume that there are constants a and b so that in any time interval of length t, the number of pedestrians arriving is $a*t$ and the number of cars arriving is $b*t$. The numbers a and b can be found by doing some experiments in which we observe the arrival of pedestrians and cars at a typical crossing. (If you want to be fancy then you can model the arrival of the pedestrians and the cars as random variables in which the expected time between arrivals of (say) a pedestrian is $1/a$. However, for simplicity we will not do this here.)

Once the traffic lights turn green and vehicle traffic starts to move, we will need to wait a time of $1/a$ before the first a pedestrian arrives and presses the button to cross. The lights then have a delay D during which cars can drive through before the pedestrians can cross again.

Step 3 For simplicity we will assume for the moment that we can ignore assumption 5 above so that all of the pedestrians wait until the traffic lights turn red, the cars stop, and then they all cross at once in a time T.

Step 4 Now we formulate the basic assumptions as mathematical formulae.

We have a complete cycle of time $1/a + D + T$, between two pressings of the button.

In this time $1 + a D$ pedestrians arrive and then cross.

In the same time b $(1/a + D)$ cars travel over the crossing.

The average total flow F of pedestrians and cars is then given by:

$$F = \frac{1 + a\,D + b\,(1/a + D)}{1/a + T + D}.$$

Step 5 We will now use this model to make some predictions.

If we take (for example) $a = \frac{1}{2}$ s^{-1}, $b = 10$ s^{-1}, and $T = 1$ s then a plot of the flow F as a function of the delay D is given in Fig. 2.2.

We can see from this model that the total flow *increases* as D *increases*. Indeed it rises from $F = (a + b)/(1 + a\,T) = 7$ when $(D = 0)$, to $F = (a + b) = 10.5$, as D tends to infinity.

We would conclude from this simple model that **it is best to make the delay D as long as possible**.

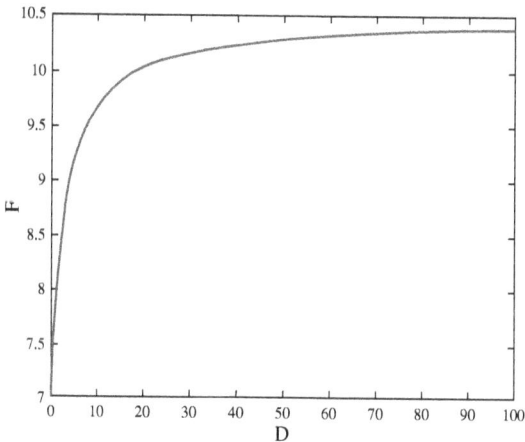

Figure 2.2   *A plot of the flow F of pedestrians and traffic as a function of the delay time D. In this graph we can see that the flow is 7 when D = 0, and tends to a limit of 10.5 as D increases.*

Step 6 In a perfect world we would now compare our model against some data. However, we don't have any available. Instead we will compare the model against our own experience of crossing the road. Whilst having an infinite value of D might be good for the motorist it won't work for the pedestrians who can't cross the road under these circumstances.

3. Making the model better

The problem, of course, is that we have ignored the 5th assumption that there is a limit to the patience of a pedestrian.

Step 7 Now let's see if we can update the model by adding a pedestrian's behaviour into the model. To do this we will assume (again for simplicity) that any pedestrian will wait for a time *at most* equal to P and then they will cross the road regardless. When they start to cross the road in this dangerous manner they will slow down the traffic.

If D is less than P then the same model applies as before.

If D is greater than P then a number $a D$ pedestrians will arrive and will cross in the time available. During this time the cars will flow freely until the first pedestrians lose their patience. This will occur at a time of $1/a + P$. After this time the cars will have to start slowing down to avoid the pedestrians who are crossing. It might be reasonable to assume that the flow of the cars drops to $b/2$ because of this. These cars will cross in a time $(D - P)$.

With this new assumption, if $D > P$ then the total number of cars which passes through the crossing is given by

$$b \left(1/a + P\right) + (b/2) \left(D - P\right).$$

The total flow is then given by the new formula

$$F = \frac{1 + aD + b \left(\frac{1}{a} + P\right) + \frac{b}{2}(D - P)}{\frac{1}{a} + D + T}.$$

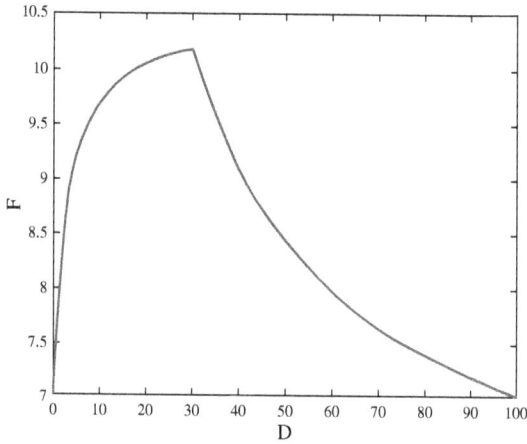

Figure 2.3   *A plot of the flow F as a function of the delay D when the pedestrians become impatient and start crossing when the traffic light is still green and the cars are still passing.*

If we now apply this formula with the same values as before, but taking $P = 30$ s (is that reasonable, ask yourselves how long you would wait?) we get a new graph for how the average flow F varies as a function of the delay D (see Fig. 2.3).

There is a big difference between this and the last graph. It has a well-defined peak (giving the maximum flow) when $D = P$. So we get the largest flow when $D = P$. This is also a safe design as in this case no pedestrian will be taking a chance when they cross the road.

Step 8 We are now in a position where we can advise policymakers. What they have to do is to make a judgement of the limits of patience of the pedestrians in their town. Then they set D to equal that. Of course this may not be an easy thing to measure.

Readers of this book might like to consider whether they agree with this model, the assumptions made to get it, or the conclusions that we have drawn from it. Indeed, how might the model be improved, and does it affect the conclusion?

For example, is the assumption that the best delay time is on which maximises the flow, or is it better to minimise the average waiting time of the pedestrians/cars? I leave these as a matter of discussion.

I suggest that you do some experiments of your own to test these assumptions when you cross the road. But do please be careful!

3
Mathematics saves the whales!

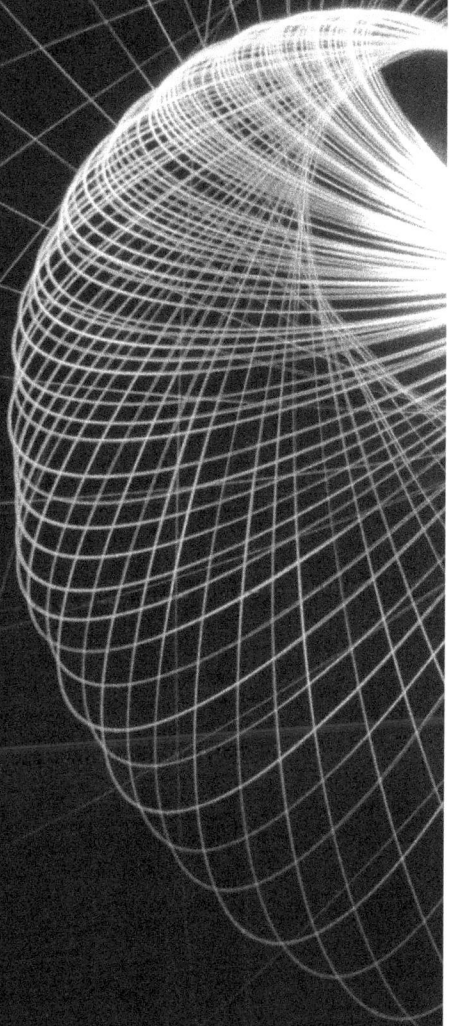

For our next example of mathematical modelling, we see how maths can save the whales. We will develop a mathematical model which involves some *geometry* and the solution of a quadratic equation. To do this I will describe some work that I have done in collaboration with Dr Philippe Blondel, a physicist at the University of Bath, who is an expert in sound propagation under the ocean, and our PhD student Dr Guillermo Jimenez Arranz.

1. Whales are under threat

It is well known that over-exploitation by the whaling industry led to serious declines in many of the world's populations of whales, although thankfully no species was brought to extinction and many are now in the process of recovering, although not all (see Fig. 3.1). However, many threats to whales remain. Many of these are due to changes in the whales' environment caused by climate change and human activity such as fishing. Others are due to the increased pollution of the oceans. Sonar, and other noise pollution, from boats, which is used to detect fish, can disrupt the guidance system of whales causing them to get lost and beach (more of this later). A final threat to whales arises from them coming too close to shipping. This can lead to collisions between the ship and the

Figure 3.1 *A sperm whale. (Credit: Gabriel Barathieu, https://upload.wikimedia.org/wikipedia/commons/b/b1/Mother_and_baby_sperm_whale.jpg)*

whale, when the whale always comes off worse. In this example we will look at how, by using some mathematical modelling we can reduce the latter risk.

2. Ship strikes and how to avoid them

Evidence of ship strikes with whales comes from a range of sources including direct reports from the vessel involved, and examination of dead whales found floating at sea or washed up on the beach. For some populations, such as the North Atlantic right whale whose main habitat is the busy waters off the east coast of the USA and Canada, the mortality rate is particularly high compared to the overall population size. It is thought that mortality due to ship strikes may make the difference between extinction and survival for this species. The most effective way to reduce collision risk is to keep whales and ships apart (see Fig. 3.2).

To make sure that a ship does not hit a whale it has to be a certain distance from it called the 'mitigation zone'. Provided that the whale is outside the mitigation zone then it will be safe. The mitigation zone varies from ship and from whale to whale, but a distance of 400 m is not unreasonable for a small ship. Thus to save the whales from ship strikes, or from surveying explosions, we must make sure that we can estimate the distance from them, so that they are outside the mitigation zone.

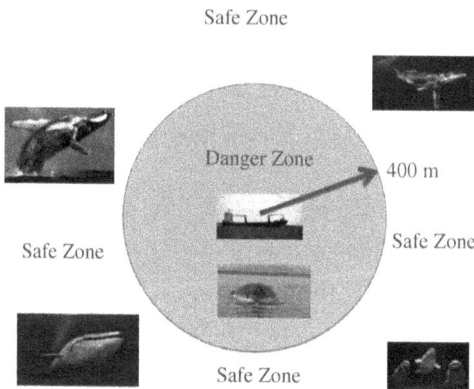

Figure 3.2   *The safe zone and dangerous zones for whales close to a ship.*

We must now work out a way of finding the distance from the whale to the ship. There are many ways to find out how far a whale is from a ship. The simplest of these is to spot it visually, and seismology prospecting ships have to carry 'whale spotters' as a result. Other techniques such as radar are less effective due to the relatively small size of the whale and its tendency to submerge. Active sonar is often used to detect fish (and submarines) and involves sending pulses of high energy sound which reflect off nearby objects. Detecting these reflections allows the location of the fish (or submarine) to be determined. However, active sonar must be avoided when searching for whales as it damages their own very sensitive navigation system and can lead them to beach and then to die. The best way by far to detect the whale is instead a passive approach where we simply listen to it. Whales 'sing' and we can detect the sounds that they make. In principle by hearing when the sounds arrive, we can work out how far away the whale is. To do this we must use a mathematical model.

3. Using a mathematical model to find the whales

We will construct our model to help us find the whales by using the same steps as before.

Step 1 We first consider the physical situation. Let the whale be a distance L from the ship. We will also assume that it is just underneath the surface of the sea, and that the sea has a (constant) depth of H. When the whale makes a sound we will hear a sound which comes directly from the whale and also one (and maybe more) which echo off from the sea bed (see Fig. 3.3). So we will expect to hear a first sound at time t_1, then a sequence of echoes at times t_2, t_3, t_4 etc.

Step 2 Now we consider the physical laws involved. Sound from the whale travels to the ship as the form of a wave which travels in the water at a speed of about $c = 1500$ metres per second. The sound frequency depends upon the species of the whale and also the type of song they are singing, but a frequency of 100 Hz is not uncommon. The wavelength of this is given by $1500/100 = 15$ m. The amplitude of the sound from the whale, satisfies the *wave equation*

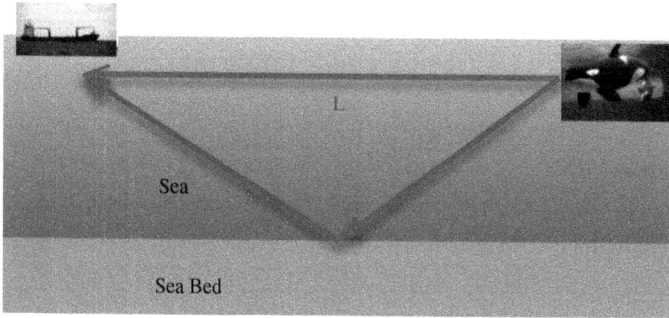

Figure 3.3   *Two paths for the sound from a whale to reach a ship a distance L away. One is the direct path. The other involves a reflection from the sea-bed.*

which is another example of a *partial differential equation.* It is possible to solve this equation directly, indeed this is exactly what oil prospectors do. However, the computer codes to do this are extremely complex and slow running. With a bit of mathematical approximation we can make the process very much faster.

Step 3 The simplification that we use relies on the observation that the wavelength of the whale song is much smaller than the 400 m distance that the wave has to travel. Because of this a good approximation is that over this distance the sound wave travels in a *straight line* just as light rays do. It is possible to fully justify this using more advanced mathematics, but we will not do this here. This is a hugely important simplification as it reduces the complex problem of wave motion, to a much simpler problem in geometry.

Step 4 Using the straight line approximation we can work out how long it takes the different sounds from the whale to arrive. If the whale is a distance L away from the boat then the time of arrival of a direct sound from the whale to the boat is given by

$$t_1 = \frac{L}{c}.$$

As an example, if the whale is 1 km from the ship then we hear its sound at the time

$$t_1 = 2/3 \, s = 666.66 \text{ ms.}$$

Now let's look at the echo of the sound of the whale. The wave for this will follow the triangular path as illustrated below in which it travels down to the sea-bed, and reflects off from it and back to the boat. If we suppose that the depth of the ocean is H and the distance travelled by the sound on its straight line path to the sea bed is S, then it will travel a total distance of $2S$ to get from the whale to the ship. We can find S by drawing a couple of right-angled triangles to represent the path of the sound wave.

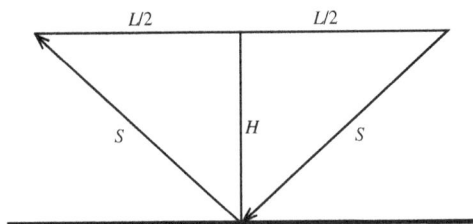

Now, we apply Pythagoras' Theorem to each of the triangles. This tells us that

$$S^2 = H^2 + \frac{L^2}{4}.$$

The total distance $2S$ travelled by the sound going from the whale to the boat via the sea bed is then given by:

$$2S = 2\sqrt{H^2 + \frac{L^2}{4}} = \sqrt{4H^2 + L^2}.$$

By dividing this distance by the sound speed c we can work out the arrival time of the echo (see Fig. 3.4). This is given by

$$t_2 = \frac{\sqrt{4H^2 + L^2}}{c}.$$

Going back to our example of the whale 1 km from the boat, if the depth of the sea is 100 m then we have the echo arrives at the time $t_2 = 679.8693$ ms.

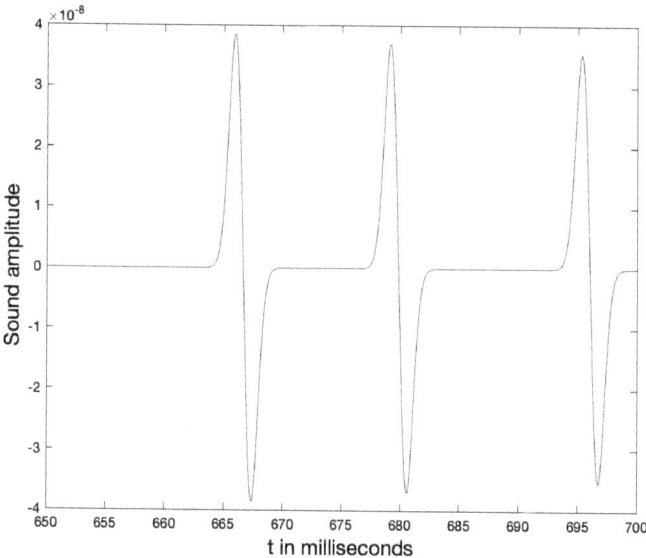

Figure 3.4 *The sequence of echoes that arrive from the sound made by the whale. The echoes arrive in pairs because one has a small extra reflection close to the ship from the sea surface.*

We show the direct pulse and the echo pulse above, which is what the ship would record (along with a third echoing pulse. It is easy from this to measure the time difference Δ between these two pulses given by:

$$\Delta = t_2 - t_1.$$

For our example we have $\Delta = 679.8693 - 666.6666 = 13.202$ ms. This is not much, but it is quite measurable on the ship's equipment.

It follows from our model that:

$$\Delta = \frac{\sqrt{4H^2 + L^2}}{c} - \frac{L}{c}.$$

Step 5 Knowing this time difference we can work out the value of L and thus find out how far away the whale is. To do this we must solve the equation above. Although it looks complicated we can work out L from it. Firstly we add L/c to both sides of the equation. Then we square both sides, which gives the expression

$$\Delta^2 + 2\Delta\frac{L}{c} + \frac{L^2}{c^2} = \frac{4H^2}{c^2} + \frac{L^2}{c^2}.$$

We can cancel the same term L^2/c^2 from each side of this equation, which simplifies things a lot. Finally rearranging we get

$$L = \frac{2H^2}{\Delta c} - \frac{\Delta c}{2}.$$

Provided that we know H then we can find L, and hence find the whale (see Fig. 3.5).

Let's check that this works with our example. We have $\Delta = 13.202$ ms. If we substitute $c = 1500$ ms^{-1}, and $H = 100$ m into the above, then you do indeed get $L = 1000$ m as advertised.

Continuing with the example of taking $H = 100$ m we can find L as a function of Δ and a graph of this calculation is given below. In this we can see that if

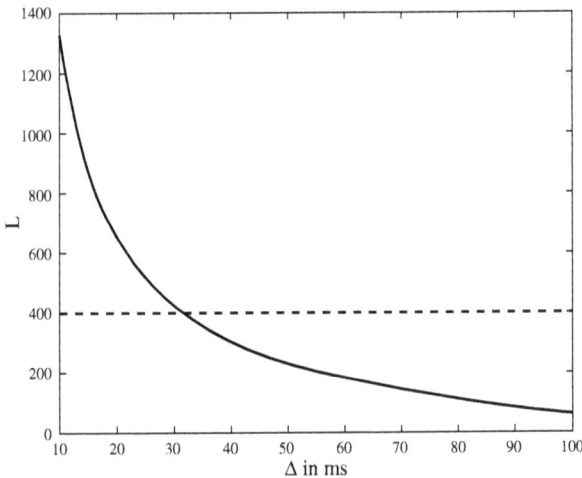

Figure 3.5 Estimating the distance L of the whale from the delay Δ between the echoes received at the ship.

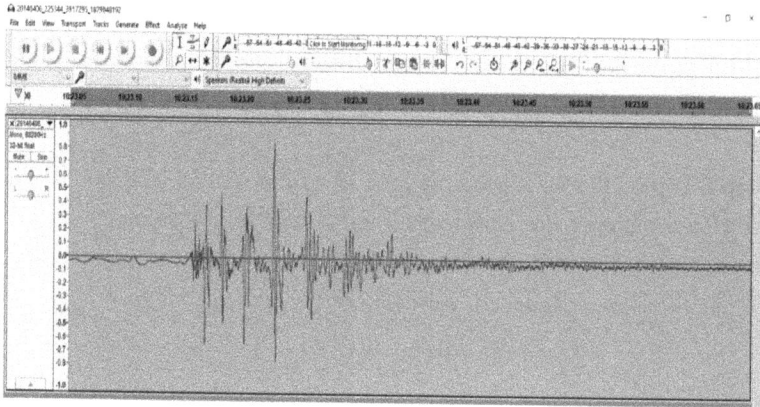

Figure 3.6 *A screen shot of an actual plot of the echoes received. (Credit: With thanks to Dr Philippe Blondel, Dr Guillermo Jimenez Arranz, and of Seiche Ltd.)*

the time difference between the arrival time of the first signal and of its echo is less than 30 ms then the whale is safe, but if it is longer than 30 ms then the whale is in trouble. By means of this simple formula we can thus save the whales.

Step 6 Figure 3.6 shows a screen shot of some real data taken from a ship by my colleagues Dr Philippe Blondel and Guillermo Jimenez Arranz, using equipment developed by Seiche Ltd. You can see that there are pulses exactly as predicted. There is additional 'noise' which you can see as an extra ripple on the signal. This is because the sound travels as a wave rather than exactly as a straight line. But the effect of this does not prevent us from being able to see the spikes clearly. You can also see that the spikes change in size. This is mainly caused by the properties of the antenna on the ship which is used to detect them. Despite all of this the sound signals are easily close enough to those predicted from our model for us to be able to find the whale.

Step 7 Of course as with all models we have made many simplifications and assumptions to get to this result. One of these is the assumption that we know the sea depth *H*. One way to find this, which is commonly used, is to employ active sonar. However, this itself can injure the whales. A much better way is to make use of the information in all of the other echoes from the sea-bed which we can see above. By listening to these it is possible not only to locate the whale, but to also find H and much more besides.

Step 8 Putting this all together we have a whale detector which works, and which is completely safe. Now to get it installed on lots of ships.

So, mathematics really does save the whales!

4

How mathematics helped in the fight against COVID-19

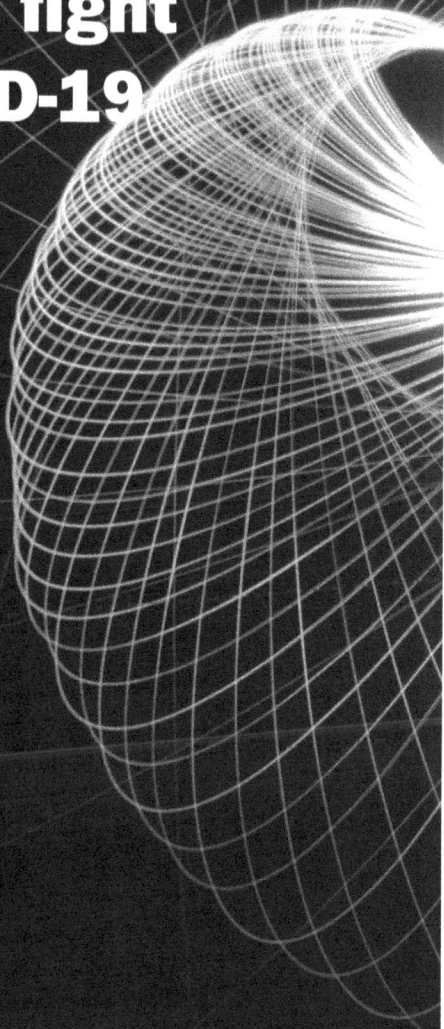

1. Overview

One of the most important applications of mathematical modelling are the models which are used to predict the spread of an infectious disease such as COVID-19 (see Fig. 4.1). During the COVID-19 epidemic such models were used widely to inform policy. They were used for many reasons, such as predicting how the disease would spread, what its impact would be on schools or on universities, and on finding the safest way to operate trains, organise shops, and running large events such as football matches. Models were used to try to understand what was happening on any particular day, what would happen in the next week, and what the long-term outlook would be. They were also used to find the best ways to moderate our behaviour so that everyone stayed as safe as possible. In doing so, the models used gained a certain degree of notoriety in the process. Mathematical modellers suddenly found themselves both on TV and also at the heart of government, asked to justify a modelling

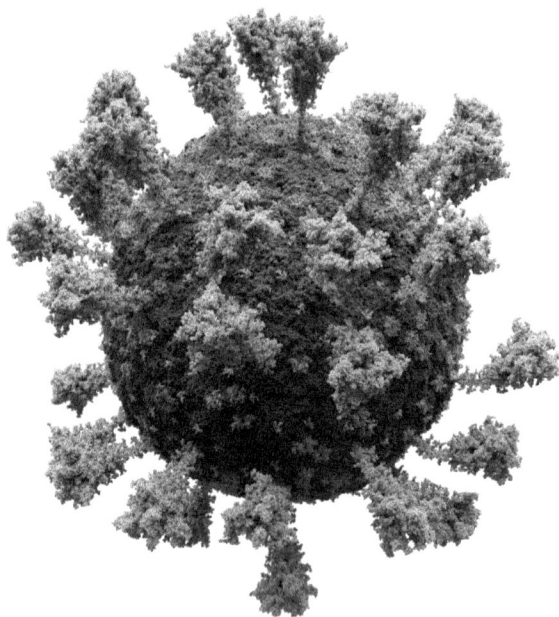

Figure 4.1 The COVID-19 virus. (Credit: Alexey Solodovnikov, Valeria Arkhipova, https://upload. wikimedia.org/wikipedia/commons/9/94/Coronavirus._SARS-CoV-2.png)

prediction. In doing this they had to consider all of the issues that we looked at in Chapter 1. In particular, how reliable the models were, could the assumptions behind them be justified, and how useful they could be to predict an inherently difficult problem which involves not only disease, but also human behaviour. The general public were treated to an insight into how the scientific process worked, with the challenge to the scientists and modellers that they were trying to make predictions about a virus which was not well understood at the start of the pandemic, with limited data. This was also the case where literally life and death decisions were made on the basis of the predictions made by the modellers. These were very challenging times for all concerned, with models being developed very rapidly, usually in parallel with gathering data on the spread of the pandemic. It is a tribute to all concerned (and to the very long, and stressful, hours that were put in by the mathematical modellers) that the predictions of these models, whilst they obviously came with huge uncertainties, were as good, and useful, as they were.

2. The SIR model for the spread of an epidemic

We will now look at how such models for the spread of an epidemic work. We will concentrate on a simple model which looks at how a disease spreads which skips over the very difficult issue of how people respond to the epidemic, and how they should moderate their behaviour. This model will involve the use of nonlinear ordinary differential equations, similar to the ones that we met in Chapter 1 but in a different context. Here we will see how differential equations can help us to save lives by giving some insight into the growth of the pandemic. Fortunately for all a lot of work had been done in advance of the COVID-19 epidemic on how diseases spread.

I should point out that the actual models used during the COVID-19 pandemic were much more sophisticated than the ones I will describe in this chapter, and they took many more factors into account. For example, much of my own work during the pandemic was involved in modelling the risk of infection in buildings (for example, a university lecture theatre or a shop) and also in trains.

In this case we had to study the way that the air flow caused the virus to spread around and the way that people would breath air which contained the virus in and out. The mathematical techniques for doing this were similar to the ones used in weather forecasting that I will describe in Chapter 6.

To understand how an infectious disease spreads from one person to another we will build up a model by using the same procedure as before. I should warn you that some of the mathematics in this Chapter is tougher than we have seen earlier. This is to be expected. Predicting the spread of a disease is a serious problem and to solve it we need to use some hard mathematics. But the results of the mathematics give important guidance about what to do in a pandemic. You may prefer to read forward to **Step 8** in Section 4 where we discuss how the conclusions of the mathematical reasoning can be, and indeed were, used to inform policymakers.

Step 1 The basic premise behind most models of the spread of an infectious disease is that if there is a population of N people, then at any time t there are a certain number I people (called the Infected population) that are infected with the disease. Whilst they are infected they can pass it on to others, and they will either recover or die from the disease. There are S people (called the Susceptible population) who can potentially catch the disease from the infected population. There are also a number R people who have either recovered from the disease, or have died, and in either case play no further role in the transmission of the disease.

Step 2 We will now try to see what sort of laws might govern the above processes. A crucial number is R_0, which is the *transmissibility* of the disease. R_0 is the total number of people that could be infected by a single individual during the time that person is infectious. If there are I infected people then the number that they infect will be $R_0 I$.

The value of R_0 depends on the disease. The larger it is the more infectious the disease is. For the usual type of flu R_0 is about 1.5. For COVID-19 R_0 is between 2 and 3 (depending upon the variant) making COVID-19 roughly twice as infectious as the flu. However, the value of R_0 for measles is about 15. This means that measles is far more infectious than COVID-19. Why then doesn't everyone get the measles? The answer is that most people have had measles at some point in their lives. They have now recovered, and having recovered they cannot catch measles again. In particular they can no longer get infected.

Important Note We have just made a huge modelling assumption. That is that the probability of someone getting infected by another person is a constant value R_0. In practice the chance depends on a multitude of other factors, such as how close people are, how they breath in the virus, how well ventilated the environment is, how old they are, whether the schools are open, etc. In this chapter we will average out all of these factors to give the single number R_0. However, the sophisticated models developed before, and during, the pandemic, tried to take all of these other factors into account as well as possible.

The proportion of the population that could get infected in the time T is the susceptible population divided by the total population or S/N. Hence, if the total population is N then (on average) the number of people infected is given by

$$R_0 I S/N.$$

Thus in this time the number of infected people *increases* by $R_0 I S/N$ and the susceptible population decreases by the same amount.

To complete this model we have also to consider that in the same time period T a proportion p of the infected population will stop being infected. This is because they (hopefully) recover, or (sadly) they die. So in the time T the total number of infected people changes by the amount

$$R_0 I S/N - p I,$$

and the number R of people who have recovered increases by the amount $p I$.

Now, all of this is constantly changing, so that people are constantly being infected and recovering from the infection. The best way to understand this is to use *differential equations* which allow us to model something which is continually changing.

To construct a *differential equation model* for the spread of COVID-19 we look at the rate of people getting the disease. To do this we set

$$\beta = R_0/T \text{ and } \gamma = p/T.$$

This might look all (or part) Greek to you, and you'd be right. Mathematicians often use Greek letters to represent important constants to avoid confusion with other variables in Roman letters. In this case we are using these two Greek letters for the important constants related to the different rates of infection and recovery. Using these letters, the differential equation model for the spread of the disease is then given by:

$$\frac{dS}{dt} = -\frac{\beta IS}{N},$$

$$\frac{dI}{dt} = \frac{\beta IS}{N} - \gamma I,$$

$$\frac{dR}{dt} = \gamma I.$$

This is the celebrated and very important **SIR model** for the spread of an infection. It is a set of *three coupled nonlinear ordinary differential equations.* Versions of the SIR model were at the heart of the very sophisticated models widely used during the COVID-19 pandemic to predict its spread.

Step 3 In the previous models at this stage we have looked at how we can simplify things. However, there is not much that we can do to simplify the SIR model without losing its predictive power. So we will leave it as it is for the moment.

Step 4 Now we will see if we can solve the SIR model. As it is nonlinear a simple solution is simply not available. But we can still make some progress by doing some more careful mathematics. If we add the first two equations together we get

$$\frac{d(S+I)}{dt} = -\gamma I.$$

Now if we take the first equation, and divide though by S and rearrange it we can express I on the righthand side of the above equation by an expression involving S and its derivative. This gives us the following.

$$\frac{d(S+I)}{dt} = \frac{N\gamma}{\beta}\frac{dS/dt}{S}.$$

So what, you might say. This equation looks a lot more complicated than the one that we started with. The reason that we do this is that we can *integrate* both sides of this expression to get rid of the derivatives. Integration is the reverse process to differentiation, and the integral of a function $f(t)$ is the area under the curve that we get when we draw the graph of f.

We write this as

$$\int f(t)\,dt.$$

An important result about integration, which was discovered by the French mathematician Pierre de Fermat (of Fermat's Last Theorem fame) is that

$$\int \frac{dt}{t} = \ln(t) + C.$$

Here $\ln(t)$ is the *natural logarithm*, which is the inverse of the exponential function so that if

$$e^x = y \qquad \text{then} \qquad x = \ln(y).$$

For example,

$$\ln(1) = 0, \qquad \ln(2) = 0.693147, \qquad \text{and} \qquad \ln(e) = 1.$$

Any scientific calculator will have a *ln* button. Try it out for yourselves. The natural logarithm function was discovered by the Scottish mathematician John Napier in 1614, and it now plays a vital role in all branches of science and engineering. The constant number C is an arbitrary constant, which we have to find later when we look at additional data about the problem.

Using Fermat's result we can integrate both sides of the expression that we derived from the SIR model. By doing this we find that there is a constant C for which

$$S + I = \frac{N\gamma}{\beta} \ln(S) + C.$$

The value of the constant C depends upon the number of infected and susceptible people at the start of the epidemic. In a real epidemic we start with I being small and S close to N (so that not many people are infected, and most of the population is able to be infected). Then to a good approximation C is given by

$$C = N - \frac{N\gamma}{\beta} \ln(N),$$

and then

$$\frac{S}{N} + \frac{I}{N} = \frac{\gamma}{\beta} \ln\left(\frac{S}{N}\right) + 1.$$

These formulae allow us to find I in terms of S. We can then substitute into the first equation to get a single differential equation for S. After a bit of rearrangement to simplify it, this is given by

$$\frac{ds}{dt} = -(\gamma \ln(s) + \beta(1-s))\, s,$$

where $s = S/N$ is the proportion of susceptible people in the population. This formula is a lot simpler than the whole SIR model, but at this point we have to stop, as there is no way to solve this differential equation exactly!

But, please do not despair! We can still learn a many more useful facts about the solutions of the SIR model, and it is also easy to solve on a computer.

Step 5 Of crucial importance to politicians and others, are firstly how rapidly the epidemic will grow at the start, and also when it will have reached its peak. Both of these are questions which we can answer using the SIR model. Here is a plot of a typical solution showing S, I and R, with an initial population of $N = 1{,}000$, of which $I = 20$ are infected at the start of the epidemic. In this plot you can see the number of infected cases rising rapidly every day, reaching a peak at about 40 days, and then declining slowly.

3. Predictions of the growth of an epidemic using the SIR model

When an epidemic starts the number of susceptible people will usually be more or less the whole population (assuming that they do not have immunity built up by exposure to other similar diseases before). This is because there has not been time to build up immunity, or to vaccinate anyone, or to put in place any sort of measures such as lockdown. This means that initially we can approximate S/N to be equal to one. The equation for I then simplifies to

$$\frac{dI}{dt} = (\beta - \gamma)I,$$

which has the solution

$$I(t) = e^{(\beta - \gamma)t} I(0).$$

This is called **exponential growth**. What we are seeing is very rapid growth with the number of cases doubling every few days. At this point the epidemic appears to be completely out of control.

If the epidemic continues this way then the number of susceptible people will drop as they become infected and then either recover or die. This means that the rate of people becoming infected will start to decrease. The number of people being infected will therefore increase at first, reach a peak, and then decrease. This is the pattern that has been seen through the ages for infectious diseases.

Here is a figure for the number of infected cases each day in the UK during the COVID-19 pandemic up till 19 December 2021. In this figure we can see a number of times when the number of cases grew exponentially (for example, when the Delta-variant took hold in late May 2021 or the Omicron variant in December 2021). The modelling challenge is to work out what will happen next.

4. Reaching the peak of the epidemic

We reach the peak of infections when I has its maximum value. A well-known result in calculus discovered by Isaac Newton tells us that this happens when $dI/dt = 0$. For the SIR model this point comes when

$$\frac{\beta SI}{N} - \gamma I = 0.$$

Dividing through by I, we find that the number of infected people reaches its peak when

$$\frac{S}{N} = \frac{\gamma}{\beta}.$$

This way we can estimate the number of susceptible individuals at the peak. If we substitute this value into the earlier formula for $S + I$, we find that at the peak the proportion I/N of infected people is given by:

$$\frac{I}{N} = \frac{\gamma}{\beta}\left[\ln\left(\frac{\gamma}{\beta}\right) - 1\right] + 1.$$

Using this formula we can accurately estimate the maximum number of people infected during the epidemic. This makes all of the previous hard work worthwhile, as the result of this is a really useful estimate that can be used to help a policymaker. As an example we can take realistic values taken from the data generated in the COVID-19 epidemic, and or which

$$\beta = 0.18 \text{ and } \gamma = 0.072.$$

These were the values used to give the plot in Fig. 4.2. For these values the formula above gives

$$I/N = 0.233.$$

This is a very large value. It means that nearly a quarter of the whole of the population becomes infected. This would have led to a huge number of fatalities if it had been allowed to happen, with predictions of 500,000 in the UK alone (see Fig. 4.3). It was this prediction which led many nations to bring in a state of lockdown.

Step 6 I said above that we used 'realistic values for the parameters from the COVID-19 epidemic'. These values were obtained by looking at how the pandemic was spreading and then fitting the predictions of the SIR model to the data that was emerging. By calculating the values at the *start* of the pandemic

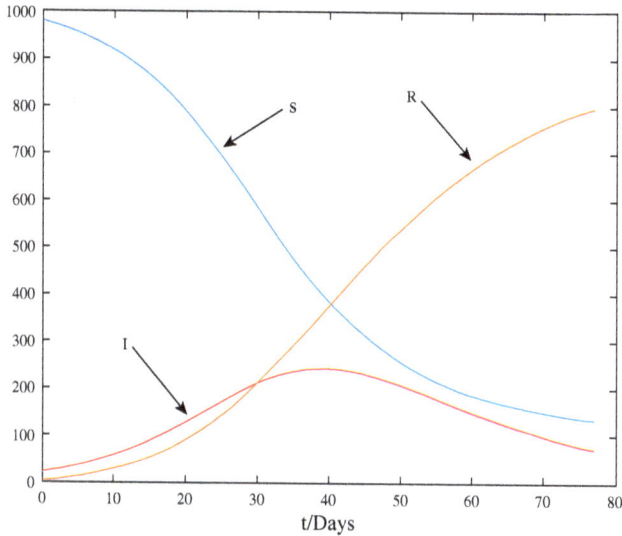

Figure 4.2 *A solution of the SIR model showing the change in the susceptible (S), Infected (I) and recovered (R) populations.*

Case numbers rising

Daily confirmed coronavirus cases by date reported

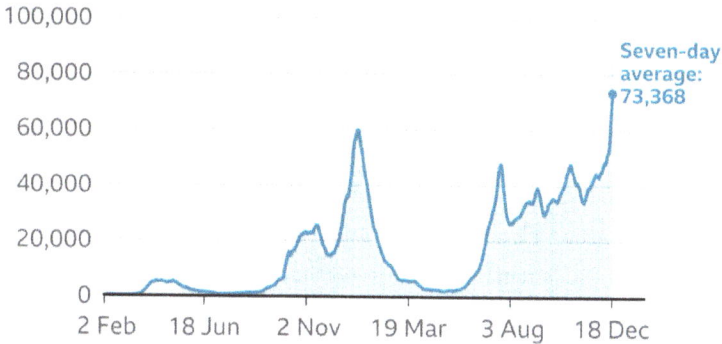

Note: Testing not readily available in first months of the pandemic. Lateral flow tests became widely used in spring 2021

Source: Gov.uk dashboard, updated 18 Dec

Figure 4.3 *BBC data for the daily number of COVID-19 cases in the UK from February 2020 till December 2021.*

it was then possible to predict how the pandemic would *continue*. This process is called *calibration* and is an example of how the model allows us to make good use of the data that we have to hand, and to then make future predictions from it.

Step 7 The SIR model whilst simple to understand and interpret only gives a very limited picture of how an epidemic spreads. In practice during the COVID-19 pandemic much more sophisticated models were used. These included including the effects of people moving around, shutting schools and universities, and the impact of keeping vital shops open. These models often had taken many years to develop, and they were constantly updated during the pandemic. An important extra was the addition of quantified uncertainty in the estimates given by the model.

Step 8 The simple result that the number of new infections is given by $I R_0 S/N$ played an important role in **establishing the policy** that governments used to control COVID-19. This shows that even simple mathematical models can be very useful.

5. Using the SIR model to inform, and guide, policy

Critical to this planning, and often reported in the media was the number

$$R = R_0 S/N.$$

If $R < 1$ then the epidemic is under control as fewer people get infected than are currently infected. Conversely if $R > 1$ then the epidemic grows exponentially out of control. Thus strenuous efforts were made to ensure that R was less than one. The key way to do this is to keep S small. This way the number of people who can get infected is also small.

There are various ways of keeping S small.

One method is to wait until a large proportion has caught the disease and have then recovered (or in extreme cases have died). This is called *herd immunity*.

Until recently this was the only real way to deal with a disease. It is certainly what happened with the Bubonic Plague in the Middle Ages, and more recently with the Spanish Flu in 1918. However, as we saw in the calculation above, with COVID-19 such a policy would have led to an unacceptable number of deaths, and was therefore not adopted.

A *second* method is to deliberately make S small by preventing people from coming into contact with anyone who might be infectious. This is the principle behind *lockdown*. Lockdowns were widely used during the COVID-19 pandemic. They do not cure any disease, and once the lockdown ends (as we found to our cost) the disease simply comes back again. However, lockdowns are effective in controlling the disease for a while, preventing the health services from being overwhelmed, and giving time for other measures to be brought into place.

The third method is to reduce S by vaccinating the population. This is the method which finally did bring COVID-19 under control (after a long period of lockdown). If we take $R = 1$ then this means that

$$S/N = 1/R_0.$$

Another way to think about this is that the proportion of the population V which has to be vaccinated is given by

$$V = 1 - S/N = 1 - 1/R_0.$$

If $R_0 = 2.5$ then $V = 0.6$. Or simply put, to keep COVID-19 under control we had to vaccinate

60% of the population.

This is a huge number, which is why the vaccination procedure took up such a lot of time.

Indeed, when the Delta-variant of COVID-19 arrived the value of R_0 was measured to be closer to $R_0 = 3.1$. In this case we get $V = 68\%$, so even more of the population had to be vaccinated.

This all took a huge amount of planning. Without the mathematical model the scale of this would never have been understood at the time and the vaccination procedure would have been far less effective.

There is no doubt that modelling played an important role in helping policymakers respond to the COVID-19 pandemic. But I think that I will leave the last word to Professor Graham Medley of SPI-M.

> *Modelling is a really useful and increasingly essential tool but it has some uncertainty, and the inherent uncertainty is that we can't predict what people will do.* [8]

I couldn't have put it better myself.

5

Can mathematics predict the future?

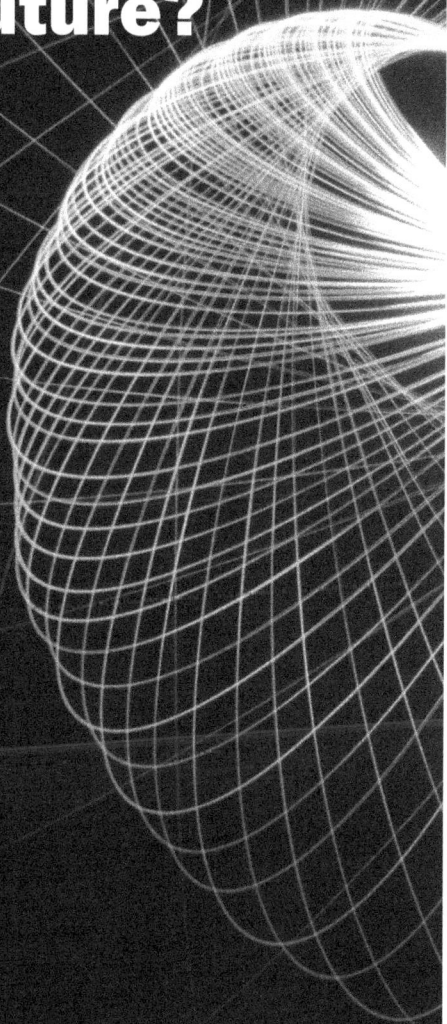

1. Introduction

In this chapter we will ask the challenging question of whether we can use a mathematical model *to predict the future!* Admittedly this is one of the most important questions we could ever ask. Whole industries (including for example gambling, and insurance), not to mention many major religions, as well as soothsayers, mystics, astrologers, governments, financial advisors, and climate scientists, owe their very existence in part to attempts to answer this very question. So, I ask again, is it possible for us to predict the future? Hands up for your thoughts on this.

In one sense there is an easy to answer. For a human being it is very likely that for most practical purposes of importance to our lives, we can predict one second into the future. Much of what we see around us on human time and length scales doesn't change much, if at all, over a period of one second. A computer model of the weather, or of the climate, or of the complexities of a city, would have little difficultly predicting the behaviour over the next second with high accuracy. However, for a sub atomic particle one second is a very long time indeed over which a great deal can happen. Computer models of particles and molecules typically operate over very short time-scales and find it hard to predict for such a long time as a second (which is why computer simulations of chemical reactions are very computationally intensive and time consuming).

If I was to ask again, can we predict what is going to happen in an hour's time, again we could be pretty sure of the answer. I expect those of you who are reading this book will still (I hope) be here, and that the world hasn't ended. Almost certainly you will still be alive (although you may not still be reading this book) and, although we will have aged slightly, your bodies will function much the same as they did an hour ago.

If we now extend this to a day we can still predict some things quite well. For example, with over 70% probability the weather tomorrow will be the same as today. This is called the *persistence forecast* and any weather forecast has to be better than this to be deemed useful. It takes a surprising amount of careful computation to do better than the persistence forecast, and I will return to this issue in the next chapter. The trains will probably run to the same time-table and my house will still be there. However, other things may have changed much more, for example the stock market can crash in a day, or a violent storm can arrive which both devastates the rail network and wrecks my house (as happened in the Great Storm of 1987!).

The further we go into the future the harder it is to say anything with certainly. For example, reliable weather forecasts are usually impossible after a period of about ten days, and no one can predict what the economy will look like in a year. Almost no one at the start of 2019 could have predicted the huge changes to our lives that happened in 2020 when COVID-19 arrived.

Indeed as the great Danish physicist Niels Bohr said,

> *It is hard to predict anything, especially about the future.*

Notably Niels Bohr was one of the fathers of quantum theory which has uncertainty at a sub-atomic level, built into its very fabric.

Central to our understanding of how well we can predict things, is explaining how the same things can change (and the associated level of risk associated with such changes). Often change is slow, as we have seen in the case of the weather. (In fact, if you live in certain parts of the world then the weather can seem to be the same every day.)

Similar slow change is seen in evolutionary systems in biology, or in geology. Sometimes change is expectedly fast, such as in a volcanic eruption, or a 200 m hurdles race. Both are, in a sense, predictable events. What is much harder to predict are sudden changes (also called tipping points or *black swan events*) which arise when a seemingly stable (one might say boring) system changes suddenly and catastrophically. This might arise through the action of a significant external cause, such as the extinction of the dinosaurs, 65 million years ago, by an asteroid or the sudden arrival of the COVID-19 virus at the end of 2019. However, it can also arise of its own accord when the accumulation of small causes suddenly leads to a very large and possibly irreversible change. Traditionally this is thought of as *the straw which broke the camel's back*.

Examples of this sort of behaviour occur in nature such as an avalanche, in engineering such as a power cut (as we shall see in Chapter 7 on Energetic Mathematics), or in human history, such as the very rapid chain of events, which led to the start of World War I. We are possibly seeing a similar type of behaviour with the current changes to our climate. Insurance companies would love to have good answers to the question of whether we can predict sudden change. However, this seems to be philosophically the contradictory question of *Can we predict the Unpredictable?*

In this chapter, (and the next one on climate change) we are going to have to look at using mathematical models not just to predict the future, but also to see how predictable the future can be, how small changes now will affect things in the future, and whether we are likely to see things going on more or less as they are now, or instead will change suddenly. In doing this we will combine mathematical modelling with a touch of chaos.

2. A brief history of telling the future

We saw in Chapter 1 that not only could a mathematical model (based on Newton's laws) predict the motion of a pendulum, but that models based on the same laws could not only predict the motions of the planets, but could even predict the existence of new planets.

These great success stories gave the mathematicians and scientists after Newton great confidence in the ability of mathematics both to describe the world that they lived in, but also to predict how it would behave in the future.

In many respects they were right to do this. Indeed the design of modern technology such as the mobile phone or an aeroplane is based on the power of mathematics (and computers based on that mathematics) to predict exactly how the phone or aeroplane will work, well before it is actually made.

In fact, a mobile phone works because of Maxwell's laws for electromagnetism. These were a set of differential equations linking electrical, **E**, and magnetic fields, **B**, together with the current **j** and charge density ρ, that were written down by James Clerk Maxwell as a mathematical model. That model summarised the empirical observations of Michael Faraday in the experiments that he did at the Royal Institution in London (at which august establishment I happen to be the current Professor of Mathematics). Maxwell's equation expressed in the current notation of vector calculus is given in the following. Technically speaking they are a set of *coupled linear partial differential equations*. Remember the word *linear* — we will come back to that later in a big way.

$$\nabla . E = \frac{\rho}{\varepsilon_0},$$
$$\nabla . B = 0,$$
$$\nabla \times E = -\frac{\partial B}{\partial t},$$
$$\nabla \times B = \mu_0 \, j + \frac{1}{c^2}\frac{\partial E}{\partial t}.$$

In these equations we will see some of the expressions that we met in Chapter 1. In particular the terms $\partial B/\partial t$ and $\partial E/\partial t$ describe how the magnetic and electrical fields change in *time*, and the terms $\nabla . E, \nabla . B$ etc. describe how they change in *space*.

Ok, rather like some of the differential equations that we met in Chapter 1, these equations look pretty scary, but that didn't put Maxwell off. By using mathematical methods which had just been developed to find the solution to partial differential equations, Maxwell *was able to solve them exactly*. What he found was remarkable. The solutions were *waves* of electricity coupled to waves of magnetism. When he calculated the speed of the waves the result was simply astonishing. The waves *had the same speed as the (also recently calculated) speed of light*. I wonder what Maxwell thought when he discovered this, for he had done nothing less than to find the formula for light, simultaneously uniting the fields of electricity, magnetism and optics. I know what I would have done! At this point Maxwell had used his mathematical model to *describe* the universe. But now he pushed it further to *predict something new*. Maxwell realised that his equations had many solutions which corresponded to different waves of different frequency and wavelength, but all moving at the speed of light. Some of these were light waves (of different colours), some were of higher frequency (such as ultra-violet waves and X-rays) and some had a slightly lower frequency (such as the infra-red rays which had also recently been discovered). But some, of lower frequency were *completely new* and a novel prediction of the mathematical model. We now call these radio waves. Not only did Maxwell's equations predict their existence (and speed) they also allowed the mathematicians of the day to predict many of their properties. They were only found experimentally, by Heinrich, some years later.

Pause to think what the world would be like without the discovery of radio waves. We would have no radio, no TV, no radar, no Wi-Fi, no Internet, no mobile phones, and no micro-wave cookers!

> *Simply put the modern world would not exist!!!*

All this is a result of the predictions of a mathematical model.

What else could be done? Indeed, it was felt in the 19th Century that as the universe was made up of atoms, and the atoms obeyed Newton's laws, then their motion could be predicted with high accuracy and for a long time and hence mathematical models could be used to predict just about anything. This led the celebrated French mathematician Pierre Simon Laplace (who made many profound contributions to both pure and applied mathematics) in 1814 (see Fig. 5.1) to make the following bold statement:

Figure 5.1   *Pierre-Simon Laplace, inventor of Laplace's equation, the Laplacian, and also of Laplace's Demon. (Credit: 1842 posthumous portrait by Madame Feytaud, courtesy of the Académie des Sciences, Paris)*

We may regard the present state of the universe as the effect of its past and the cause of its future. An intellect which at a certain moment would know all forces that set nature in motion, and all positions of all items of which nature is composed, if this intellect were also vast enough to submit these data to analysis, it would embrace in a single formula the movements of the greatest bodies of the universe and those of the tiniest atom; for such an intellect nothing would be uncertain and the future just like the past would be present before its eyes.

This intellect is often called 'Laplace's Demon' [9] and it is the ultimate mathematical modeller. It seems to be able to know the future exactly, and there seems also to be no chance, unpredictability or free will. The possible existence of such a demon is in itself a very frightening prediction. Another way of putting this, which was very popular for a time, was that we live in a *clockwork universe* which 'ticks' in the same predictable way as a mechanical clock.

But is it really true? Indeed, realistically how good are mathematical models at predicting the future, and are mathematical modellers really demons?

3. Where does it all go wrong? Enter chaos theory

Unpredictability in nature

If we think about our reality for just a moment, it is hard to equate Laplace's bold prediction and the existence of his demon with the world that we observe around us. Perhaps for the human condition, this is no bad thing. Despite the work of Newton and Maxwell, much of the world really does appear to be unpredictable, especially where human beings are involved.

Examples of unpredictability in nature and in human behaviour occur everywhere from the behaviour of the Financial Times Stock Exchange (FTSE) index (see Fig. 5.2) to that of my dog, Monty. Indeed human behaviour is inherently unpredictable, and we are able (in defiance of Laplace's Demon) to exercise free will.

Unpredictably also occurs in the physical world even when we are quitesure of the underlying physical laws. Despite the existence of the Navier–Stokes equations which give an excellent model of the way that the atmosphere moves around and are excellent for short-term predictability of this (hence the success of the daily weather forecast), they are hopeless in telling us what the weather will be doing in ten days. Similarly certain climatic phenomena are also very difficult to predict. A good example of this is the El Nino Southern Oscillation (ENSO) phenomenon in which we see a rapid warming of the Southern Atlantic Ocean roughly once every four years. A plot of the Southern Ocean temperature

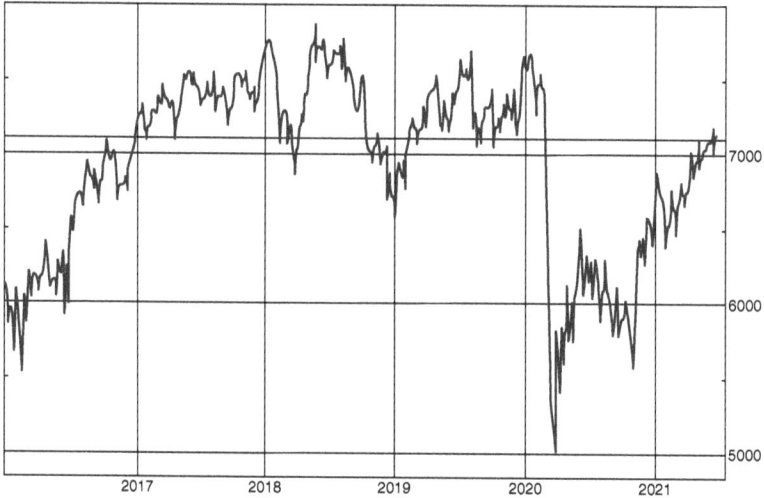

Figure 5.2 *The FTSE index over the last five years. From day to day it can appear very irregular. Note the big crash in 2020 due to the COVID-19 pandemic.*

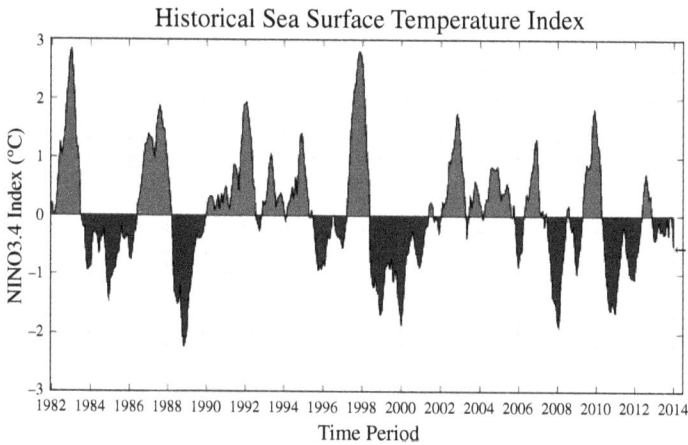

Figure 5.3 *The erratic behaviour of the temperature in the South Pacific Ocean. Roughly every four years there is a warming in the El Nino effect. It is still hard to predict exactly when this will occur.*

is given in Fig. 5.3. Whilst there is a rough periodicity of four years to the ENSO, the plot below the sea surface temperature (which it is believed is mainly caused by large-scale ocean circulation patterns) is anything but regular. We shall return to the ability (or not) of mathematics to predict the climate in the next chapter.

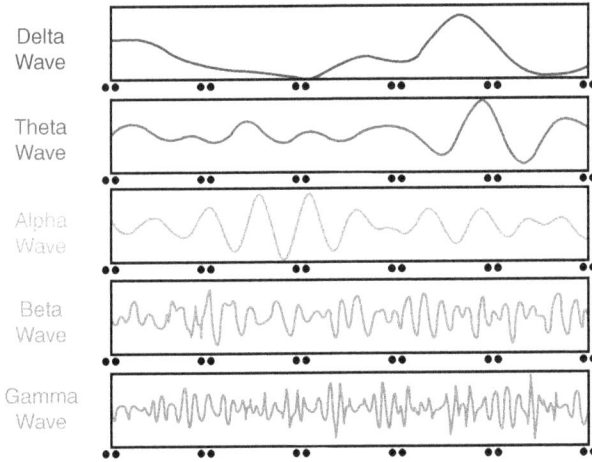

Figure 5.4   *Seemingly erratic behaviour in brain wave patterns.*

Seemingly erratic behaviour can also be observed in various brain wave patterns as measured by an EEG. These are shown in Fig. 5.4.

So, unpredictability really does seem to be all around us, from the motion of molecules to that of the atmosphere, and from the behaviour of a dog to the waves in the brain.

However, this seems in contradiction to the ordered clockwork universe predicted by Laplace and others. With the success of Newton's laws in predicting the future in some cases we are thus led to ask the following question:

> *Does the unpredictability we see in many cases in nature, arise because nature really is complex and unexplainable.*

Or in contrast

> *Can seemingly unpredictable behaviour arise from systems governed by Newton's (and other physical) laws?*

We will now try to answer this difficult question.

Chaos and the double pendulum

Before we rush into the complexities of the climate (in the next chapter) we will try to answer this question by looking at a relatively simple system, which, without question, is governed by Newton's laws of motion. This is the *double pendulum* that is an extension of the simple pendulum studied by Galileo which we looked at in Chapter 1, and which opened our eyes to the power of mathematical modelling.

A double pendulum is two pendulums coupled together, in much the same way that our leg has two bones coupled by the knee joint. Each of the two pendulums is free to move independently under the action of gravity, but they are joined together by the pivot in the middle. Here is an example of a double pendulum (and various other mathematical gadgets) in my own office (see Fig. 5.5).

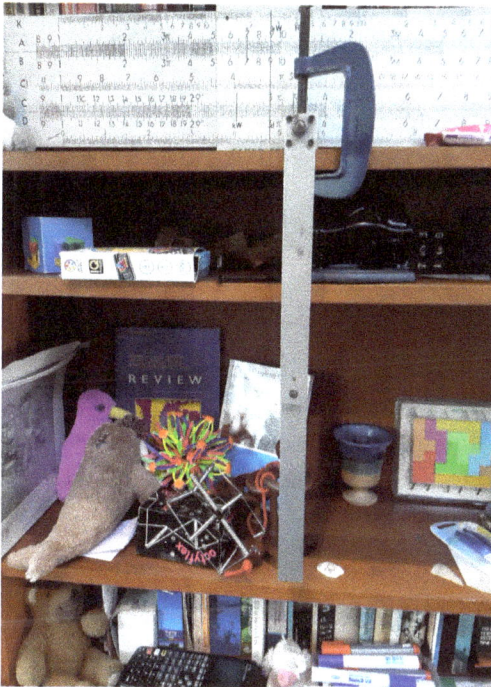

Figure 5.5   *The double pendulum in my office. This very device appeared in the 2005 Royal Institution Christmas Lectures given by Marcus du Sautoy. Behind it is a giant slide rule.*

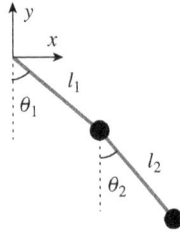

Figure 5.6 *A schematic of the double pendulum.*

A schematic of the double pendulum is shown in Fig. 5.6. In this we will think of the double pendulum as made up of two freely swinging bars of length l_1 and l_2. The top bar swings from a (frictionless) pivot at the top, and the second from a (frictionless) pivot on the first pendulum. The bars make respective angles of θ_1 and θ_2 to the vertical.

What could be simpler. This system has only two moving parts (the top and bottom parts of the pendulum), each of which has a position and an angular velocity. It has thus only *four degrees of freedom.*

This is far fewer than the weather, which has billions of degrees of freedom. We might be tempted to think that the double pendulum will therefore have nice simple behaviour.

> *We would be wrong!!!*

Just like the weather the double pendulum can have remarkable complex behaviour. In fact it has three distinct types of motion, two of which are simple and predictable and the third of which is complex and anything but predictable.

If the top and bottom parts of the pendulum are pulled together a *small* amount to one side, then they will continue to swing to and from *in phase* (meaning that they move in the same direction) in a regular manner. See the left part of Fig. 5.7. This is just like the periodic motion of the simple pendulum and is very predictable.

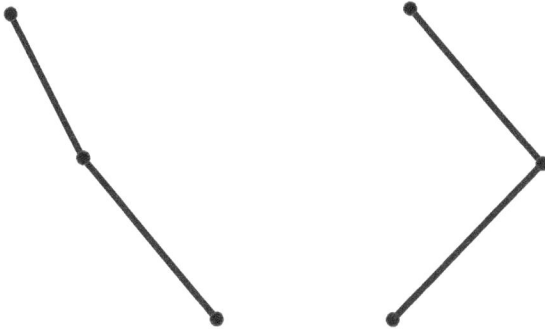

Figure 5.7 *Two types of regular periodic motion of the double pendulum. On the left the motions of the top and bottom section are in-phase, on the right the motions are exactly out-of-phase.*

In contrast, if the two pieces are pulled in *small* opposite directions and released then they will continue to move periodically to and fro but in opposite directions. This is what is called *out-of-phase* motion. See the right part of Fig. 5.7. This periodic motion will also continue indefinitely in a very predictable way.

Finally, but most interestingly, we will give the pendulum a large swing and see what happens next. What we see is truly remarkable. Instead of periodic motion, the double pendulum will continue to move *in a most erratic and seemingly random manner.*

You can find many examples of this on the Internet. Just type 'double pendulum' into Google and watch the movies. See if you can predict what will happen next.

An illustration of this is given in Fig. 5.8. In this picture a light is attached to the lowest part of the double pendulum and a picture of the motion of this recorded over time. Not only is the motion extremely complex and hard to predict, if the pendulum is set off again as close as possible to the same position as before, then after a very short time it will move in a completely different way. Such a pendulum seems to defy the predictability we observed earlier and would be a poor time-keeper!

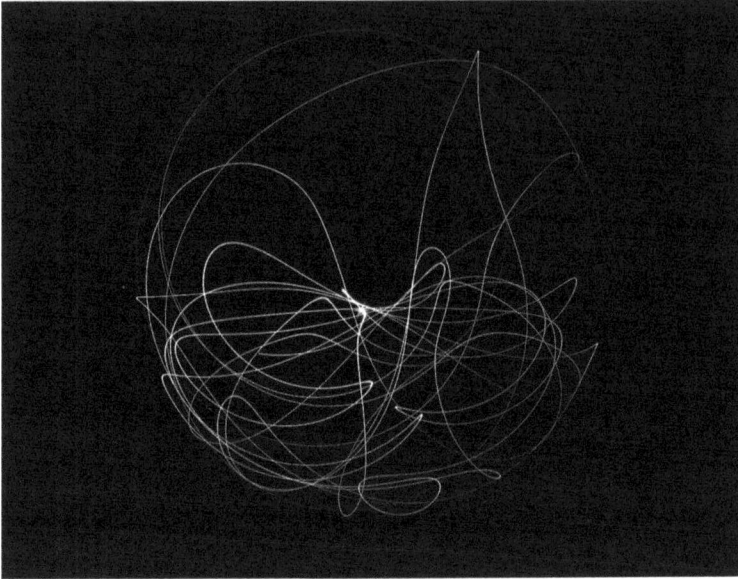

Figure 5.8 *The chaotic motion of the double pendulum. In this figure we follow the motion of a light on the bottom of the lower arm of the pendulum. (Credit: Cristian V., https://commons.wikimedia.org/ wiki/Category:Double_pendulums#/media/File:Chaos_Theory_&_Double_Pendulum_-_4.jpg)*

> *We call this motion chaotic.*

Chaotic motion has two main features:

> - *It is extremely complex. It does not seem to obey any predictable pattern.*
> - *It displays 'sensitivity to initial conditions'. If you start a chaotic system off in two slightly different ways then the two types of motion rapidly become completely different.*

If this resembles the way the weather behaves, then that is no coincidence. Both the pendulum and the weather obey the same physical laws of motion. It is just that the weather has more degrees of freedom than the pendulum.

So, what is going on here? It might be possible to argue that chaotic motion appears to be random because the double pendulum is simply responding to random air currents which are blowing it around in lots of different directions. However, this is not the case. The air currents certainly have an effect, but it is not them that make the motion of the double pendulum chaotic.

To show this we will create a mathematical model of the double pendulum's behaviour.

We will follow the same procedure that worked so well for the simple pendulum by writing down the equations of motion of the double pendulum and then solving them.

We start by thinking about what is going on. In its usual form, the double pendulum (a little unlike the case of the chandelier we looked at in Chapter 1) comprises two rigid metal bars of respective masses m_1 and m_2, and lengths l_1 and l_2. The top bar is pivoted on a fixed, frictionless pivot, and makes an angle θ_1 to the vertical. The lower bar is pivoted at the lowest point of the upper bar (again frictionlessly) and makes an angle θ_2 to the vertical. The whole system is then acted upon by gravity. If the masses m_1 and m_2 are reasonably large then, to a good approximation, we can ignore the resistance of the air. We will also ignore any vibrations in the top pivot (in practice we have to clamp the double pendulum pretty hard to make sure that this doesn't).

Now we write down the equations. I will make a confession here. This is not easy. To do this you need to use quite advanced mathematical techniques which involve calculating the Lagrangian of the double pendulum (which is the difference between its kinetic and potential energy) and then applying Lagrange's equations of motion. (Alongside Laplace, Lagrange was another famous French mathematician who's name began with L. There are at least two others, Legendre and Lebesgue.) Usually these techniques would only be taught in the advanced mechanics course of an undergraduate degree. However, they are very powerful, and essential to master if you want a career in robotics or satellite design or indeed in any field where you want to work out the advanced motion of a coupled system. This includes computer animation, and the movie industry. (Advert: Hollywood employs lots of mathematicians for exactly this reason.)

To cut a long story short the motion of the double pendulum can be *completely described* by the following pair of *coupled nonlinear second order ordinary differential equations* which give us our mathematical model of its behaviour. A complete derivation (and much other great material about dynamics and chaos) can be found in the excellent book by David Acheson [19]. These equations are similar to those for the single pendulum that we met in Chapter 1, but there are two of them because we have two different parts of the pendulum.

$$(m_1 + m_2)l_1\ddot{\theta}_1 + m_2l_2\ddot{\theta}_2 \cos(\theta_1 - \theta_2) + m_2l_2\dot{\theta}_2^2 \sin(\theta_1 - \theta_2)$$
$$+ g(m_1 + m_2)\sin(\theta_1) = 0,$$
$$m_2l_2\ddot{\theta}_2 + m_2l_1\ddot{\theta}_1 \cos(\theta_1 - \theta_2) - m_2l_1\dot{\theta}_1^2 \sin(\theta_1 - \theta_2) + m_2g\sin(\theta_2) = 0.$$

Important Note In this equation, to save space I have used Isaac Newton's notation for a derivative. In this notation $\dot{\theta} = d\theta/dt$ and $\ddot{\theta} = d^2\theta/dt^2$. Newton's notation gives nice compact formulae, but is generally harder to work with, and less transparent than the other notation for a derivative, which was developed by the German mathematician Gottfried Leibniz at the same time. There was a major punch-up at the time between Newton and Leibniz about who had invented calculus first. The answer, of course, was that neither did. Both mathematicians (in their own admission) relied heavily on earlier work by others in Europe and India, all of whom could claim a role in the discovery of calculus.

But here (as expected) is the bad/good news.

The differential equation for the double pendulum is another example of a problem that we cannot solve exactly by hand (although it is relatively easy to solve by computer).

When we encountered this issue with the simple pendulum we decided to simplify the system by only looking at small swings. This led to a linear differential equation which we could solve. We can do exactly the same thing here. If the angles are *small* then the equations describing the double pendulum system can be approximated by a set of linear ordinary differential equations by using the approximations that for small angles $\sin(x) = x$, $\cos(x) = 1$. This then gives the simplified linear system:

$$(m_1 + m_2)l_1\ddot{\theta}_1 + m_2 l_2\ddot{\theta}_2 + g(m_1 + m_2)\theta_1 = 0,$$
$$m_2 l_1\ddot{\theta}_1 + m_2 l_2\ddot{\theta}_2 + m_2 g\theta_2 = 0.$$

These two coupled differential equation *can* be solved exactly. There are two solutions, both of which are periodic. They have exactly the in-phase and out-of-phase behaviour described above.

However, we have just committed a modelling crime (that of being a mathematical drunkard). By making the 'small angle' simplification we have missed out on the really interesting behaviour of the double pendulum. (As we remarked in Chapter 1, we have to be constantly on our guard to avoid making the mistake in any mathematical model of over simplifying it to the extent that we throw away the real essence of the problem that we are trying to solve.)

If we were doing our mathematical modelling 100 years ago then that would be as far as we could get. But now with the availability not only of fast computers, but also great algorithms to solve differential equations, then we can find a very good numerical approximation to the solution of our mathematical model, and simulate its motion on the computer. If you have access to the computer language Python which has inbuilt tools to solve differential equations then this is very easy to do.

Here is example of a computer simulation of the double pendulum (using Python) in which $m_1 = 1$ kg, $m_2 = \frac{1}{2}$ kg, $l_1 = l_2 = 30$ cm and $g = 9.81$ metres per second per second. In this simulation we plot the location of the bottom of the lower part of the pendulum, and start the pendulum off with a good push.

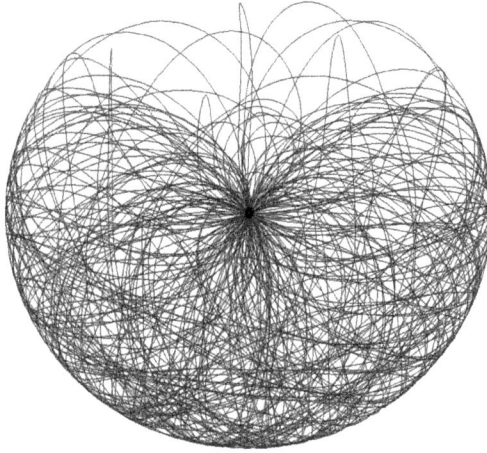

Figure 5.9 *A computer simulation using Python of the chaotic motion of the lower part of the double pendulum. This simulation should be compared with Fig. 5.8.*

Remarkably, if we compare Fig. 5.9 with Fig. 5.8 of the real pendulum we see that the computer simulation gives qualitatively exactly the same behaviour as the physical system. In particular this figure shows that the lower part of the pendulum moves almost randomly, just as in the real situation. Reassuringly the same computer simulation gives the in-phase and out-of-phase periodic behaviour when we start the double pendulum off with a small swing. In short the model works! And we can use it to make predictions, such as the average position of the pendulum (which we will see in the next chapter is relevant to climate simulations). However, what we cannot do is to make predictions of what the pendulum will be doing at a point in the future. To show you why not, we will take two very similar pendulums, which start from rest. In the first we start with $\theta_1 = 2$ and $\theta_2 = 1$ (both measured in Radians). In the second we start with $\theta_1 = 2.001$ and $\theta_2 = 1$, and we run the system for 60s. The two pendulums have started very close to each other (far closer than it would easy to arrange in an experiment). However, at the end of this time the first pendulum has $\theta_1 = -1.6466$, and $\theta_2 = 0.4249$, whilst for the other pendulum we have the very different values of $\theta_1 = -1.1372$ and $\theta_2 = -0.8078$.

What this modelling exercise has demonstrated is something very profound. The mathematical equations that we have simulated are derived directly from

Newton's laws of motion, and their solution appears to be erratic and almost completely random and unpredictable. In other words

> *Chaotic behaviour really does exist as a solution of Newton's equations.*

Or in other words, we can expect to see what appears to be random and unpredictable behaviour from systems governed by relatively simple mathematical equations.

The implications on our ability to predict the future using mathematical models is profound indeed.

There are other types of pendulums, which display similar behaviour to the double pendulum. A beautiful example is the *chaotic water pendulum* (see Fig. 5.10) which can be found in St Mary Redcliffe Church in Bristol, UK and

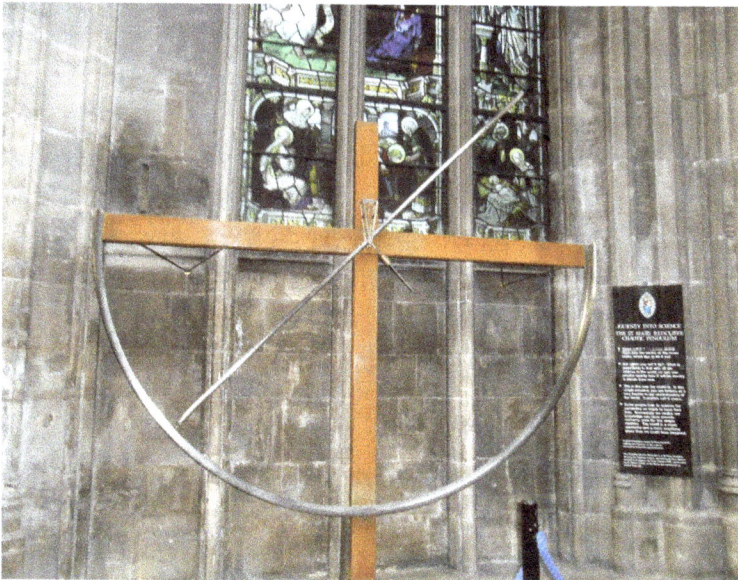

Figure 5.10 *The chaotic water pendulum at St Mary Redcliffe Church, Bristol. (Credit: Darlith Rolin, https://upload.wikimedia.org/wikipedia/commons/2/25/St_Mary_Redcliffe_Chaotic_Pendulum%2C_Bristol.jpg)*

was designed by Sir Brian Pippard FRS. Water running down the horizontal arm of the pendulum combined with the action of gravity causes the pendulum to swing in a chaotic manner. This combination of science and mystery makes this pendulum a very suitable object for contemplation in a church. Do take a look if you are ever in Bristol.

Motivated by this, from now on I will extend the earlier definition of chaotic behaviour to the following:

> *Chaotic motion is complex, irregular and otherwise unpredictable behaviour with sensitivity to initial conditions, which arises from a system which can be exactly described by 'simple' mathematical laws.*

In the popular press the sensitivity of chaotic systems to their initial conditions is often called the *Butterfly Effect*. This term was coined by Edward Norton Lorenz in the 1960s [10]. Lorenz was one of the pioneers of 'chaos theory' and he remarked that if the weather was chaotic then the flap of a butterfly's wings in Brazil could cause a hurricane in Europe. (We now know that whilst the weather is almost certainly chaotic in the sense described above, the perturbations have to be much larger than a butterfly to make any significant difference. In essence the butterfly would have to have wings about 1 km across to cause any real effect.) The idea of the butterfly effect has certainly caught the public imagination. The idea of even tiny changes leading to huge effects in the future seems to resonate with some of our perceptions of how the universe might work. Many books and even a film [11] have been written about it, and a remarkably prescient early story about it (also involving butterflies) was written by Ray Bradbury before Lorenz had even coined the phrase [12].

In a sense the butterfly effect allows us to combat Laplace's Demon. The demon in question would have to know the positions of all of the particles to infinite precision to be able to make any predictions at all. Furthermore, the inherent microscopic randomness ever present due to the action of quantum effects will be immediately amplified by the butterfly effect to make long-term predictability of a chaotic system quite impossible.

Chaotic behaviour in the manner described above exists in many, many, other physical systems. One of my favourite examples is *chaotic billiards*. Imagine a billiard ball bouncing around a table, so that when it comes into contact with the edge of the table it reflects from it in the same way that light would reflect from a mirror. The resulting pattern of bounces from a stadium shaped billiard table is given in Fig. 5.11.

The observed pattern of motion is highly complex, and yet, like the double pendulum, it arises from very simple laws of motion. This problem has very practical application in optics (to find light illumination patterns), acoustics and also in high frequency Wi-Fi. In the latter example the lines in the above picture correspond to the rays of electromagnetic radiation which bounce around a room after leaving a Wi-Fi transmitter. The very complex picture above means that it is remarkably hard to predict the strength of Wi-Fi coverage inside a domestic room.

Truly chaotic behaviour is everywhere, from the weather to the motion of the asteroids, and from the double pendulum to the behaviour of electrical circuits. Indeed it was the mathematical study of the equations of a particular type of electrical circuit (the van der Pol oscillator) by Dame Mary Cartwright during WWII which led to one of the original descriptions of the butterfly

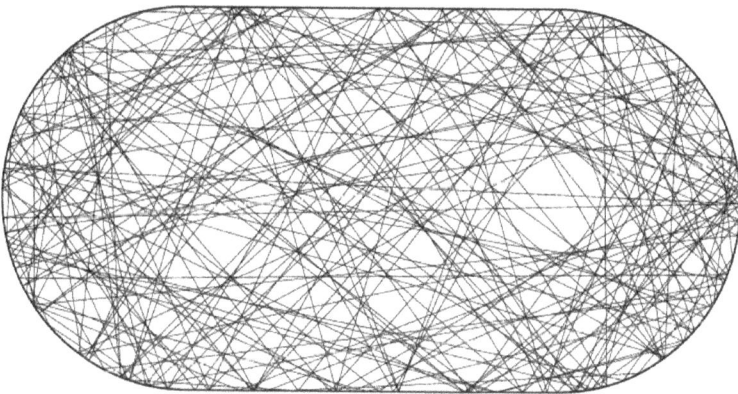

Figure 5.11 *The chaotic pattern of a ball bouncing off from the walls of a stadium. These are the same patterns as we would see with a Wi-Fi signal bouncing off the walls of a house. This explains why Wi-Fi reception patterns in a house can be so variable.*

effect. (The engineers at the time thought that the erratic behaviour that they were observing in their oscillators was due to faulty equipment. Cartwright showed that this wasn't the case, and led the engineers to better ways of designing their oscillators to eliminate the chaotic 'noise'.)

It is interesting to contrast this erratic behaviour of an electrical system, with the enormous predictability of Maxwell's equations. This leads us to ask three questions.

Firstly: What is it about some problems which means that they are likely to have chaotic behaviour.

Secondly: Can we predict *anything* about the behaviour of a chaotic system.

And

Thirdly: What is the 'use' of chaos?

The answer to the first question lies in the issue of linearity which I said was the key to the predictability of Maxwell's equations. The answer to the second is lots! But we have to be careful about what we try to predict. For the answer to the third, read the rest of this book, although I will give a short summary at the end of this chapter.

Let's look a bit further to find the answers.

4. Chaos theory

Overview, and the Lorenz equations

Although what we would now call chaotic behaviour had been observed earlier by Cartwright, the modern terms of *chaos* and of *chaos theory* arose from a paper in 1963 by the meteorologist Edward Norton Lorenz, which was published in 1963 [10]. It is no coincidence that his paper was written shortly after the availability of fast electronic computers, the existence of which made the study of chaos in detail possible. Lorenz was trying to study the motion of the atmosphere, in particular by looking at the way that hot air rises during convection (which is known to be a play a major contribution to the weather).

Lorenz followed the now, hopefully familiar, process of mathematical modelling described in Chapter 1. In particular he wrote down a system of equations for convection which were based on the Navier–Stokes equations for atmospheric motion which we also met in Chapter 1. These were much too hard to solve, either analytically, or by the computers of 1963. So Lorenz had to simplify them to make progress. He did this by assuming that there were *three* different modes (likely shapes) of the behaviour of convection, which had amplitudes $x(t), y(t)$ and $z(t)$, and which evolved in time. He then projected the convection equations onto these three modes to get equations for x, y and z. Unlike the equations for the double pendulum, the Lorenz system really is a big approximation to reality as in reality there are very many more than three modes which make up convection. By doing the projection in [10] Lorenz came up with what are now known as the Lorenz Equations. These have the form

$$\frac{dx}{dt} = \sigma\left(y - x\right),$$

$$\frac{dy}{dt} = x\left(\rho - z\right) - y,$$

$$\frac{dz}{dt} = xy - \beta z.$$

In case these equations look familiar, they have a similar form to the SIR equations which we used in Chapter 4 to study the spread of the COVID-19 pandemic. Both are systems of three ordinary differential equations, and both involve products of terms in their right hand sides. However, there the similarity ends. Whilst we can exactly predict the behaviour of the solutions of the SIR equations over long times, those of the Lorenz equations are chaotic and unpredictable.

Before the 1960s it would have been impossible to solve this system accurately. However, the 1960s saw the first use of fast digital computers, which could solve differential equations using (for example) the *Runge–Kutta* algorithm. Lorenz applied this to the equations above and was remarkably surprised. Rather than the periodic behaviour he was expecting, the solutions of the Lorenz equations were very complex. Looking for a way to describe them Lorenz coined the word *chaos* for the motion, and the *butterfly effect* for its sensitive dependence. Significantly he felt that what he was seeing reproduced the erratic and turbulent behaviour observed in the atmosphere itself. In this he was wrong, whilst the motion of the atmosphere is chaotic in a sense, it is not described in any meaningful way by the Lorenz equations.

A plot of the evolution of $x(t)$ using a fixed set of the parameters with

$$\sigma = 1, \quad \beta = \frac{8}{3}, \quad \rho = 28,$$

(which are of meteorological significance) is given in Fig. 5.12. This plot shows the chaotic behaviour with a complex trajectory as time increases. In this plot

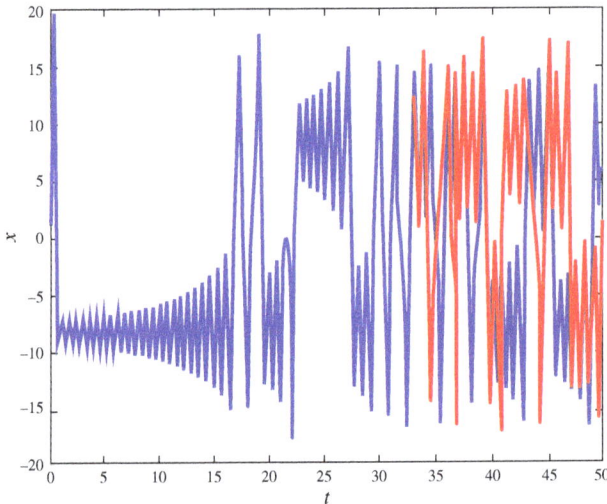

Figure 5.12 *Two solutions of the Lorenz equations computed using Python. The blue solution has x(0) = 1, and the red solution has x(0) = 1.00001. The two solutions deviate significantly after t = 32, showing the sensitivity to initial conditions of the Lorenz equations.*

two slightly different initial conditions of $x(0) = 1$ and $x(0) = 1.00001$ are taken, with $y(0) = z(0) = 0$. Whilst the two trajectories are close up to a time $t = 32$, they differ significantly after this time.

A plot of $x(t)$ and $y(t)$ together is even more revealing. In this case the points (x,y) move around an (appropriately) butterfly shaped set. This set is called a *strange attractor* because it attracts all trajectories but is neither periodic or a fixed point (see Fig. 5.13).

Although the points move around the strange attractor in a chaotic manner, the shape of the attractor is itself well defined and remains the same regardless of how the system is started off. Here we see the organisation which is at the heart of chaotic systems. Although their individual motions can't be predicted, the shape of the set that they move on can, to high precision. This allows a lot of predictions to be made about the average properties of a trajectory. This important feature of any chaotic system becomes very important in the next

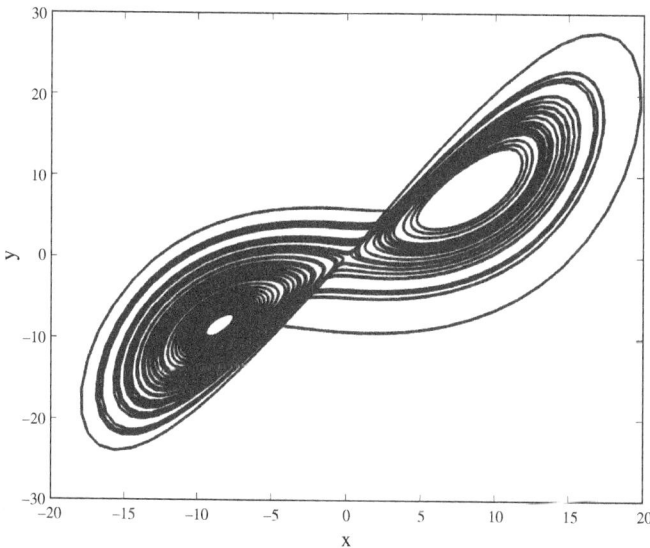

Figure 5.13 *The butterfly shaped strange attractor of the Lorenz equations computed using Python. Although the solutions of the Lorenz equations are very sensitive to the initial conditions, the shape of the strange attractor is not.*

chapter when we look at climate. The strange attractor itself has a fine structure and beauty and is an example of a fractal set. This means that if you magnify any part of the curve then it continues to have complex structures at any scale.

The discovery of chaos in the 1960s caused a sensation and led to an explosion both of research and also of articles in the popular press and a huge amount of hype, only some of which was justified. The celebrated book by James Gleick [13] made the discovery almost look like a murder mystery and was instrumental in popularising the subject of chaos theory. More scientific, but equally readable accounts of chaos theory and its implications can be found in the books by Ian Stewart [14] or Steve Strogatz [15].

As we have seen, much of the real credit for the mathematical analysis of chaos should be given to Mary Cartwright. However, in many ways the existence of chaotic dynamics had been discovered rather earlier at the end of the 19th Century, and the credit for its discovery is largely due to the great French mathematician Henri Poincaré (1854–1912) (see Fig. 5.14).

Poincaré was at the time investigating the stability of the Solar System. It was well known (as we have already seen) that if a single planet moves around the Sun, then its motion is periodic and can be predicted exactly using Newton's laws. However, Poincaré was able to show that a system of three bodies of similar mass and moving under gravitational attraction only would move in an irregular orbit. We now recognise this as just another example of chaotic motion. I will talk more about this in the final chapter of this book when 'Mathematics Goes into Space'.

The logistic map and its implications

It is hard to see how the chaotic behaviour in the Lorenz system, or the double pendulum arises. So we will now look at a simpler system, which has similar behaviour and can be studied in much more detail. This is the celebrated *Logistic Map*. Let's suppose that we want to predict the population of (say) a town from one year to next. This is a subject of great interest to a town planner who has to work out (for example) how many schools to build to meet a need in five

Figure 5.14　*The mathematician Henri Poincaré (1854–1912). Father of dynamical systems and one of the first observers of chaos. (Credit: https://upload.wikimedia.org/wikipedia/commons/f/f4/ PSM_V82_D416_Henri_Poincare.png)*

or ten years time. To make this precise we will let $x(n)$ be the population of the town in the year n, with $n = 0$ set to be the current year at which point we know the town's population $x(0)$.

A simple mathematical model of population growth was proposed in 1798 by Thomas Malthus in his book, *An Essay on the Principles of Population*. Malthus postulated that in any one year a fixed proportion of the population would be born, and an equally fixed proportion would die. This meant that the population in year $n + 1$ would be proportional to the population in the year n, or in mathematical form

$$x(n+1) = a\, x(n).$$

Here a is the constant of proportionality. This is an example of a *discrete dynamical system* in which we have a rule to advance the system from one generation to the next. In this case Malthus' model has the simple and predictable solution given by:

$$x(n) = a^n x(0).$$

If $a < 1$ then the population decreases, if $a = 1$ it stays the same, and if $a > 1$ it grows without bound, so called Malthusian, or **exponential**, growth. In Chapter 4 we saw exactly the same type of growth in the early stages of an epidemic (where R_0 was used instead of a to describe the growth rate).

Malthus himself recognised that such behaviour was unrealistic, and that the population would eventually run out of resources and then start to decline. (Just as in an epidemic you finally run out of people that can be infected.) One way to do a model is to introduce an upper limit M to the population to account for the effect of limited resources, and then to modify the Malthusian model to give

$$x(n+1) = a \, x(n) \big(M - x(n) \big).$$

If we scale $x(n)$ by dividing it by M, so that the maximum population is 1, and set $r = a \, M$ to be the *scaled growth rate* we arrive at following famous model for population growth.

$$x(n+1) = r \, x(n)(1 - x(n)).$$

We call this model for population growth a *map* as it describes the map from $x(n)$ to $x(n + 1)$. The above map is so important that it is called the *Logistic Map*. Given a value of $x(n)$ we apply the map above to it to give $x(n + 1)$. If we apply it to $x(n + 1)$ then we get $x(n + 2)$ etc. The Logistic Map is nonlinear and this is the reason why it generates such interesting behaviour. Unlike the double pendulum or the Lorenz equations, this system can be studied very easily is you have access to a programmable pocket calculator or to a spreadsheet such as Excel. I recommend that you take $x(0) = \frac{1}{2}$, program the above function in, and simply iterate forward. Experiment by seeing what happens for a variety of values of the growth rate r, taking r between 0 and 4.

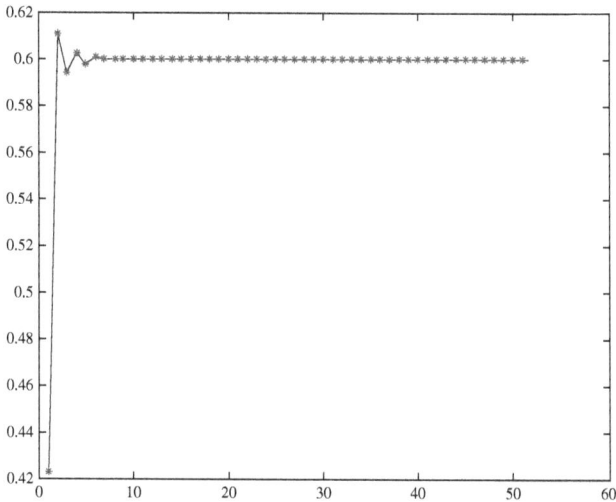

Figure 5.15   *Convergence of the solution of the Logistic map to a fixed point when r = 2.5.*

The result of your experiments will be to find that the type of behaviour that you get depends critically on the value of r.

The plot of the iterates for $r = 2.5$ is given in Fig. 5.15. From a random initial start the system evolves to a fixed final state, or a *fixed point*, x. This is the solution of the equation $x = 2.5 \, x \, (1 - x)$ so that $x = 0.6$.

Such a final state is very predictable, and would be a town planners dream. Such predictable behaviour exists for $1 < r < 3$.

When $r = 3$ something remarkable happens. The fixed point becomes unstable and is replaced by a *two-cycle* in which the population varies between two different values. (This is rather like a boom-bust phenomenon in economics and is reminiscent of the out-of-phase response of the double pendulum.) A picture of the two-cycle for $r = 3.2$ is given in Fig. 5.16.

If we increase r further to $r = 3.55$, this picture changes again, and the two-cycle is replaced by a 4-cycle, in which x varies between four values as can be seen in Fig. 5.17.

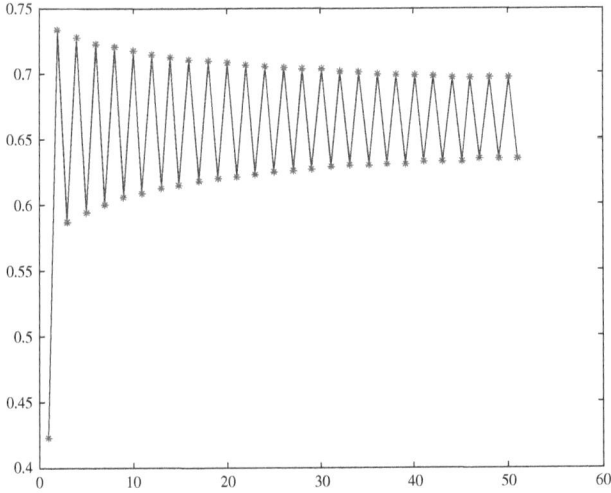

Figure 5.16 *Convergence of the solution of the Logistic map to a two-cycle when r = 3.2.*

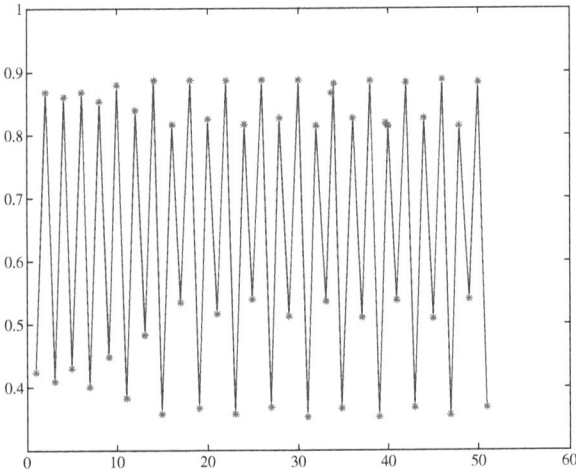

Figure 5.17 *Convergence of the solution of the Logistic map to a 4-cycle when r = 3.55.*

Increasing r leads to ever more complex behaviour, and for $r > 3.56995$ most values of r give chaotic behaviour. An example of this is given in Fig. 5.18.

It is worth reminding ourselves that this complex picture is obtained from the iterations of a very simple quadratic map.

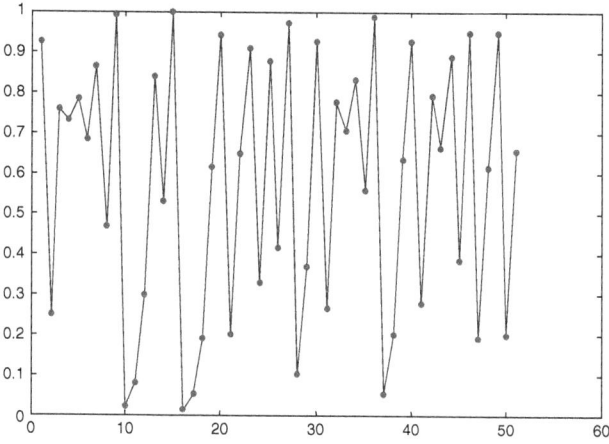

Figure 5.18 *A chaotic solution of the Logistic map when r = 4.*

We can study the case of $r = 4$ in some detail. Suppose that we set

$$x(n) = \frac{1}{2}\left(1 - \cos\left(\theta(n)\right)\right).$$

Then some simple trigonometry shows us that

$$x(n+1) = 4x(n)\left(1 - x(n)\right) = \frac{1}{2}\left(1 - \cos\left(2\theta(n)\right)\right),$$

which means that

$$\theta(n+1) = 2\theta(n).$$

We can deduce from these two expressions the following exact solution of the Logistic Map with $r = 4$:

$$x(n) = \frac{1}{2}\left(1 - \cos\left(2^n\theta(0)\right)\right).$$

Here we can see why we get chaos. Let's suppose that we take two values of $x(0)$ given by y and z which differ by one part in a million or $1/1{,}000{,}000$. The corresponding values of $\theta(0)$ will differ by a similar amount. However, every time we apply the map, the difference in the two values of θ will double. If we iterate the map 20 times then this difference gets multiplied by 2 raised to the power of 20, which is close to a million. Therefore, the difference in the two values of $\theta(20)$ for the starts at y and z will be one. The corresponding values of $x(20)$ will be very different as a result. This has shown the extreme sensitivity to initial conditions that the Logistic Map has when $r = 4$. Another way of saying this is that small changes in x are approximately doubled at each iteration, but the cosine function stops these from becoming unbounded. This leads to random-looking behaviour.

A complete picture of the behaviour of the logistic map for $2.4 < r < 4$ is given by the *bifurcation diagram* in Fig. 5.19, which plots the values of the final state of $x(n)$ as a function of r.

Note The word *bifurcation* is used by mathematicians to describe a point where the behaviour of a system changes qualitatively. In this figure a bifurcation occurs at $r = 3$ when we change from having a single fixed point, to oscillating between to different points. You can see many other such bifurcations in Fig. 5.19.

In Fig. 5.19 we can clearly see the single fixed point for $r < 3$, the period doubling to a two-cycle at $r = 3$, and the chaotic behaviour for $r > 3.56995$ including $r = 4$. There is also a stable 3-cycle around the value of $r = 3.828$. Much mathematical effort has been put into understanding this diagram in detail.

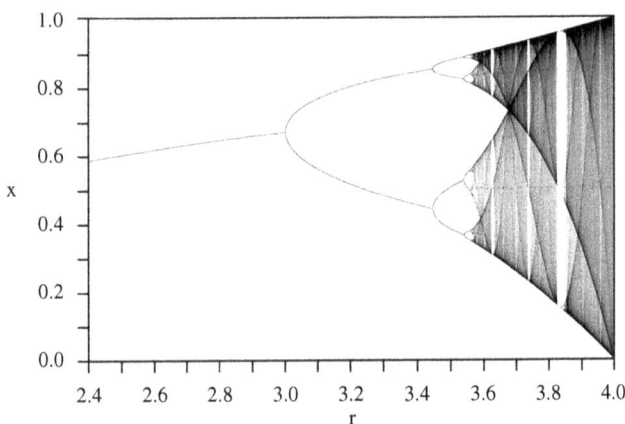

Figure 5.19   *A bifurcation diagram showing how the final solution to the Logistic map changes as r varies.*

Dynamical systems theory

We have looked at a number of examples of complex dynamics including the double pendulum, the Lorenz equations and the Logistic Map. All of them seem to have similar types of behaviour. The mathematical theory of *dynamical systems* (see Strogatz [15]) gives a way of understanding all of this. A dynamical system is simply something which *evolves in time* and is described by a state vector $x(t)$. This can either be a function of continuous time (as in the double pendulum or the Lorenz equations), or of discrete time as in the Logistic Map. Usually a dynamical system depends on many other parameters. In such a dynamical system there will be an initial state $x(0)$. This is then allowed to evolve. Typically the motion has a transient phase, which is dominated by the initial state, and then after a longer time patterns begin to emerge. Much of dynamical systems theory is an attempt to classify these long-term evolutionary states and also to understand their stability. As the parameters in the dynamical system vary so the long-term states change. A state can be created, it can disappear, it can lose its stability or it can change into a different state. We have seen all of this happen in the case of the Logistic Map. Typical long-term behaviours include fixed points (as in the case of $r < 3$), periodic motions (as in the case of $r = 3.2$ and also for the simple pendulum), quasi-periodic motions (a combination of periodic motions, seen for example in the tides), and of course chaotic motion. The transitions between these states can be shown to take a limited number of forms, and have been studied in great detail. They are explained by 'bifurcation theory' (which is a development of 'catastrophe theory' that was very much in vogue in the 1970s). The bifurcation diagram for the Logistic Map above displays many of these features, including the celebrated 'period-doubling route to chaos'. An excellent account of this theory is given in Phillip Drazin's book [16].

5. Tipping points where things change quickly

In the introduction to this chapter I talked about the example of the straw which broke the camel's back. What is happening in this case is that the camel (or rather its back) is the solution of a dynamical system. The parameter which controls the behaviour of this system is the amount of straw on the back of the camel. For small amounts of straw the camel's back is a *stable* fixed point of this dynamical system. The camel is able to bear the load and it will remain in its loaded state for a long time into the future. However, at a critical value of the amount of straw, the fixed point becomes *unstable*. The effect of this is that the camel's back breaks, and the camel and straw collapse onto the ground, possibly never to rise again.

What we have seen here is a *tipping point* where a small change in the parameters governing the system leads to a *large irreversible change* in the resulting state of that system. As part of our ability to predict the behaviour of such systems we need to be able to identify whether they are close to, or maybe have even passed through, a tipping point. Electrical engineers trying to avoid a power cut due to an overloading of the power supply network, are acutely interested in such tipping points. We will look at the causes of power cuts in Chapter 7. Studying tipping points is also of huge importance to our ability to predict the future climate and we will look at this topic in depth in Chapter 6 when we consider the factors leading to rapid changes in climate.

From the perspective of dynamical systems theory, the simplest model for a tipping point comes from the quadratic ordinary differential equation:

$$\frac{dx}{dt} = r + x^2.$$

In this dynamical system x represents its state, and r is the parameter which controls the state. As always, t is time.

If $r < 0$ this dynamical system has two *fixed points* at $x = y$ and $x = z$ where

$$y = -\sqrt{-r}, \quad z = \sqrt{-r}.$$

Of these two points the lowest y is *stable*. If you start close to y then you get attracted to y as time increases. Once you get to the point y then you stay there for evermore. This behaviour is of course very predictable.

However, the point z is *unstable* and if you perturb the system a little bit then it will evolve into a completely different state. If you start *below z* then the system evolves to y and if you start *above z* then the system goes off to infinity.

However, if the load is increased so that $r > 0$ then there are *no fixed points* and x increases without bounds and is *always unstable*.

We illustrate this in Fig. 5.20, where the arrows show the evolution of x with time t, and the fixed points y and z are plotted.

The point $x = 0$ when $r = 0$ is called a *tipping point*. (It also has the name of a saddle-node or a fold bifurcation.) Imagine that you start at the stable fixed point y and then slowly change r from being positive to being negative. For the negative values the system stays at the slowly changing value of y.

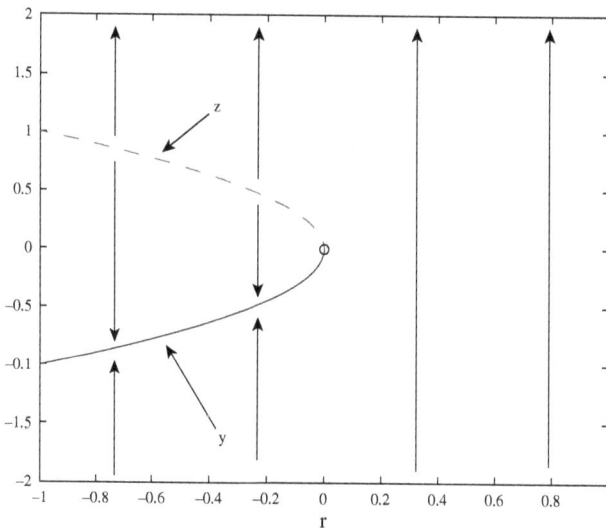

Figure 5.20 *The parabolic curve shows the locus of the fixed points y and z, as a function of r. If r < 0 there are stable (y-solid line) and unstable (z-dashed line) fixed points. At r = 0 these coalesce at a tipping point, and there are no fixed points if r > 0. The arrows show the directions of motion of the system as it evolves.*

However, when we hit $r = 0$ then everything changes and we shoot off to infinity. Everything changes at this value, hence the name tipping point (in allusion to an apple cart suddenly tipping over). Historians and economists call such behaviour a *black swan event.*

In the context of camels, the state x would represent the position of the camel's back and r would represent the load on the camel. At a critical value of the load (corresponding to $r = 0$) the load is too much and the camel collapses.

I will return to the implications of this for climate change in Chapter 6. If you are interested about what happens at tipping points in a huge number of applications have a look at Ian Gladwell's book [17].

6. Is chaos theory useful?

You bet it is! In many different ways. Many mathematical theories start out as being seemingly abstract and of no possible use, and then later become central to science and technology. Dynamical systems theory, including chaos theory, is a very good example of this.

Chaos theory in the real world

Whilst Lorenz's work in the 1960s was largely theoretical, it was quickly realised that many physical systems, such as the double pendulum, really did behave chaotically. Many other important systems are now understood to be chaotic. These include the electronic oscillators studied by Mary Cartwright, the weather (over periods of about two weeks), car exhausts, car suspensions, tubes in a boiler, power supply systems, friction brakes, climate change, solar coronal mass emissions, Wi-Fi, EEG signals, ECG signals, and the motion of the asteroids, to name just a few examples. Chaos theory allows us to understand, measure and, in some cases, control, the uncertain behaviour that such chaotic systems generally exhibit. We now appreciate that chaotic behaviour is part of the natural scheme of things for anything governed by complex, nonlinear, deterministic processes. A case in point is the asteroids. Rather than having simple orbits, many of them have highly complex orbits. It is vital that we understand this as otherwise we may not be able to predict when or if an asteroid may hit the Earth, wiping out all life in the process. Thus chaos theory is of critical importance in saving the human race!

Fractals

In a rather less terrifying application, chaos theory, and to be precise the theory of fractals is playing an increasingly useful role in computer graphics. We have seen in the example of the Logistic Map, how a simple rule can generate a graph with great complexity. If we apply the same process to (two-dimensional) pictures, it is possible for a simple rule acting on a two-dimensional (phase) space to generate a picture of remarkable complexity. There are hints that nature uses similar processes, and hence the patterns so produced often look very realistic. As an example Fig. 5.21 shows the Barnsley Fern Fractal, which has been produced by exactly this process.

Roughly speaking a fractal is a pattern produced by a simple rule which has great complexity, so that it doesn't matter how much you expand any part of the fractal, you will continue to see more and more detail.

Two very famous examples of fractals arising in mathematics are the Julia set and the Mandelbrot set (Fig. 5.22). These are both closely linked to the Logistic Map that we looked at earlier, but in this case considered as a map from one *complex* number to another.

Since their discovery these fractals have been celebrated as examples of exquisite mathematical beauty and they can be found on posters and coffee cups everywhere.

Fractals also occur in nature, for example in the snowflakes that we looked at in Chapter 1, in the shape of the clouds that we will look at in Chapter 6, and

Figure 5.21 *The 'Barnsley Fractal'. This is generated by repeatedly iterating a simple two-dimensional map. This produces a pattern that resembles a fern. (Credit: Kimbar, https://upload.wikimedia.org/ wikipedia/commons/6/69/Bransleys_fern.png)*

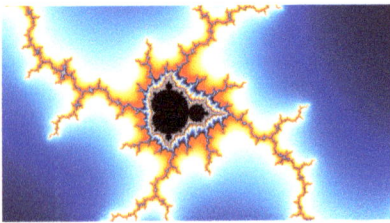

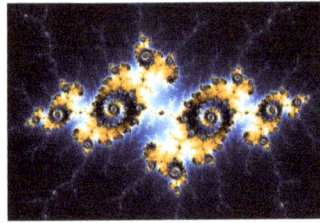

Figure 5.22 *Sections of the Mandlebrot set. These have a beautiful fractal form. (Credits: Left: https:// www.shutterstock.com/image-illustration/this-fractal-discivered-by-mandelbrot-xx-1677191488. Right: Binette228, https://upload.wikimedia.org/wikipedia/commons/a/a7/Mandelbrot_set_image.png)*

Figure 5.23   *Coastlines (such as that of Norway) often have a fractal structure. This makes it very hard to calculate the length of a coastline. (Credit: Allice Hunter, https://commons.wikimedia.org/wiki/File:Regions_of_Norway_by_HDI_(2017).svg)*

in the shape of coastlines, such as that of Norway (and for which Slartibartfast was given an award in *The Hitchhiker's Guide to the Galaxy*) (see Fig. 5.23).

A complete account of the theory of fractals can be found in K. Falconer's book [18]. I urge you to learn more about this wonderful subject.

But perhaps the most important aspect of chaos theory is how it helps us understand how predictable the world is. We will now have a look at this.

7. So, can we predict the future at all?

What chaos theory tells us is that whilst some systems (those with stable fixed points or periodic orbits) can be very predictable, others are not. What we mean by that is that even given huge amounts of computing power, and an exact knowledge of the chaotic system in question, there would always be a time after which any prediction of the exact state of the system would be impossible. Laplace's Demon is not all-seeing after all.

However, we can still predict many things about a system even if it is chaotic. For example, I cannot tell you what the weather will be on any particular day in a future year, but I can predict that it is very likely that it will be hotter on a summers day than on a winters day.

Because all chaotic systems are described underneath by simple laws, it is often still possible to make predictions about what the system will do *on average* and often much more. (Mathematicians call such long time averages the *ergodic properties* of the system.) For studying the climate this is far more important than knowing what the weather will be on any particular day. Hence, we can still predict the climate, even if we can't predict the weather far into the future. So Laplace's Demon still has some of the answers up their sleeve.

As an example, we will return to the Logistic Map with $r = 4$, which we have shown to be chaotic. Suppose that we iterate this map 200,000 times. This will generate 200,000 iterates $x(n)$ which will look like random numbers. If we plot a histogram of these numbers (with 40 bins) then the pattern we get has a lot of regularity (see Fig. 5.24).

In fact with a bit of mathematical reasoning we can show that if y is chosen (randomly) from the set of the iterates then the probability P that $y < x$ is exactly given by the expression

$$P = \frac{1}{2} + \frac{1}{\pi} \arcsin(2x - 1).$$

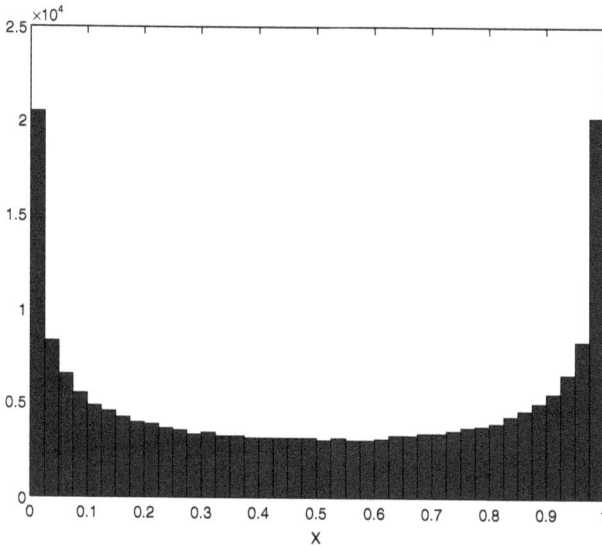

Figure 5.24 *The histogram of the iterates of the Logistic Map when r = 4. Although the iterates themselves are unpredictable, the histogram has a very predictable shape.*

Using this expression we can tell a great deal about the iterates. For example, they have an average of ½ and a standard deviation of 0.3536.

Similarly we will be able to deduce a lot about the ergodic properties of the climate. We will look at this in the next chapter.

What next?

It has been my privilege to work in this fascinating field for most of my career. I regard chaos theory as a vital way of understanding the world with almost limitless applications. In the next chapters we will see how this understanding helps in predicting the future of our climate, controlling our power supply, and designing new materials.

6

The mathematics of climate change

1. Introduction

An overview

Without question, climate change, and its effects on the environment and on society, is one of the most important, and controversial, issues facing all of us at the current time. It raises high passions on all sides of the media and of the political spectrum. Whilst the vast majority of scientists believe that there is good evidence for human influenced climate change, there is by no means 100% agreement as to either its importance or of the impact in the future. Hugely important questions remain such as. Is climate change happening? If it is happening is it due to human or natural causes? Will the effect of climate change be positive or negative? If there are negative effects and they are due to human causes, can anything be done about them, or are we past the point of no return?

In this chapter we will be concentrating on the mathematical aspects of these questions using both the ideas of mathematical modelling described in Chapter 1 and the dynamical systems and chaos theory we met in Chapter 5. We will combine these with the other areas of probability, statistics, and scientific computing, to clarify our current understanding of the climate. Using these we can make predictions for the future climate, and can qualify these predictions by clear measures of their uncertainty. An important reason for doing this, is that many of the current predictions, which are used by bodies such as the Intergovernmental Panel for Climate Change (the IPCC), are based on huge computer models. Indeed the IPCC reports, for example [20], are based on these predictions. It is these models which have also led to the 2016 Paris Agreement [21], that we should restrict our CO_2 emissions in such a way to keep the Earth's temperature rise to below 1.5°C. These complex models are, in turn, based on mathematical formulations of the physics governing the climate, informed by statistical measurements of the existing climate. Thus to have an informed debate about the future of the climate of this planet, the public has a right to know how these models are constructed, how they are tested, what sort of predictions they make, and

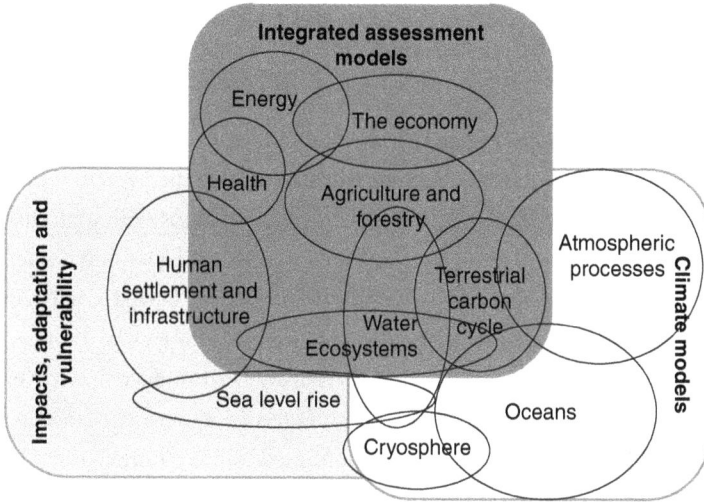

Figure 6.1 *The many components of a full Earth System Model (ESM).*

(crucially) how reliable are their predictions? In fact climate models are the most certain (or the least uncertain) of a whole set of models used to determine the effect of the climate on human beings. I illustrate these in Fig. 6.1. For this chapter I will talk about the right hand part of this figure, which also includes the variations in the amount of energy that the Earth receives from the Sun.

Without question weather has a huge affect on us all, and if weather patterns are changing due to climate variation then we need to take this seriously. I cannot resist telling you a story from my own experience. A few years ago I organised a seminar at my home institution of the University of Bath. The seminar was to have been given by an expert from the Met Office and her subject was "The effect of extreme weather on the transport network". On the morning of the seminar she rang me up. Owing to the effect of severe weather on the transport network the trains between Exeter and Bath were not running, so she could not come! How I enjoyed sending the email explaining why the seminar was cancelled (and the wonderful responses that I received), it made the point better than actually giving the seminar itself.

(I should say that later in the year we did have the seminar, and it was excellent.)

I will start this chapter by looking at evidence for climate change, both in the past and in the present. I will then describe the way that mathematical climate models work and show how these can make predictions with quantifiable uncertainty. Then I will finish by asking the mathematical question of whether the climate has reached one of the tipping points that we looked at in Chapter 5. Anyone interested in the broader issues of environmental change, its implications on society, and the changes that we need to make in view of it, can find a very readable account of the issues related to climate change in the Ladybird book titled *Climate Change*, authored (amongst others) by Prince Charles (now King) [22].

A simple mathematical model of climate change

Before we get into the details of climate change, I will set the scene by looking at a simple, but very useful, mathematical model which allows us to make predictions of the Earth's temperature. (We will look at more detailed models later, including a more careful derivation of the model I am about to present.)

This model, which was derived by the Swedish scientist Svante Arrhenius in the late 19th Century, is based on the simple observation that the Earth receives short wave energy from the Sun. Some of this energy is reflected back into space, and the rest warms the Earth up to a temperature T. The Earth then radiates this energy back into space as infra-red radiation. Some of this radiation heats up the Earth's atmosphere, which in turn re-radiates it back to the Earth, leading to additional heating. The more Carbon Dioxide there is in the Earth's atmosphere, the less transparent it becomes to the infra-red radiation, and therefore more radiation gets re-radiated back to the Earth's surface and the surface becomes warmer. This is called the 'greenhouse effect'. Simply put, this means that the more Carbon Dioxide there is in the atmosphere, the hotter the Earth becomes. This is why we are seeing global warming as a result of Carbon Dioxide produced by human activity.

Mathematically we can represent this using the following model (with full details given in Section 4). If the solar radiation is S, then the average amount reaching the Earth is

$$(1 - a)S/4,$$

where a is the albedo of the Earth (which is a measure of how much radiation is reflected back into space). The amount of this energy radiated back into space is then given by

$$e\sigma T^4.$$

Here σ is a constant (the Stefan–Boltzmann constant) and e is called the emissivity, which is a measure of how much of the Earth's radiation is transmitted through the atmosphere, and the temperature T is measured in degrees Kelvin. Balancing these two expressions we get a formula for the Earth's temperature given by

$$T = \left(\frac{(1-a)S}{4\sigma e} \right)^{1/4}.$$

If you substitute in the measured values of:

$$S = 1368 \text{ Wm}^{-2}, \quad a = 0.31, \quad \sigma = 5.7 \times 10^{-8} \text{ Wm}^{-2}\text{K}^{-4}, \quad e = 0.605.$$

Then you get

$$T = 288 \text{ K or } T = 15°C.$$

Which is about right for the average temperature of the Earth.

> *For an exercise, try setting e = 1 to get the average temperature of the Moon. The answer will be revealed later in this chapter.*

One of the great things about the above model is that we can use it to make predictions. Most significantly, it is well known that as the Carbon Dioxide levels in the atmosphere increase, then the emissivity decreases. The model then immediately predicts that the mean temperature increases. Hence, we experience global warming.

2. What is the evidence for climate change?

Current changes in the climate

There is a lot of statistical evidence that the current climate is changing, even if the reasons for this change and the significance of it, are the subject of hot debate. This evidence can, in turn, be used as a test of our climate change models. I will now look at four examples of this taken from the physical world, which can be measured directly and also checked in the climate models I will describe later.

Global warming

According to the IPCC 6th Assessment Report WG1 — Science Basis *"Warming of the climate system is unequivocal, and since the 1950s, many of the observed changes are unprecedented over decades to millennia"*. [23]

So, let's look at the actual evidence for this. Recent records on the Earth's temperature come from a variety of different sources including weather stations on the Earth's surface, satellites orbiting the Earth, and buoys and ships in the ocean. In the case of the weather stations, these records have been gathered reliably since the foundation of the Met Office in 1850. A graph of these is given in Fig. 6.2 using data up to 2020 from Berkeley Earth and the Hadley Centre, in which we show the Earth's average temperature each year, compared to the 1850–1900 average temperature.

In this period the average temperature of the Earth has gone up and down from year to year, due to such factors as the El Nino (a warming of the southern Pacific Ocean due to the effect of ocean currents), other ocean current related effects, and also large volcanic eruptions such as Krakatoa in 1883. However, these variations are superimposed on a trend, which is clearly rising. Indeed in 2015 the Earth was on average 1°C warmer than it was when these records began, and the last three decades have been the warmest ever recorded.

Average temperature anomaly, Global
Global average land-sea temperature anomaly relative to the 1961–1990 average temperature.

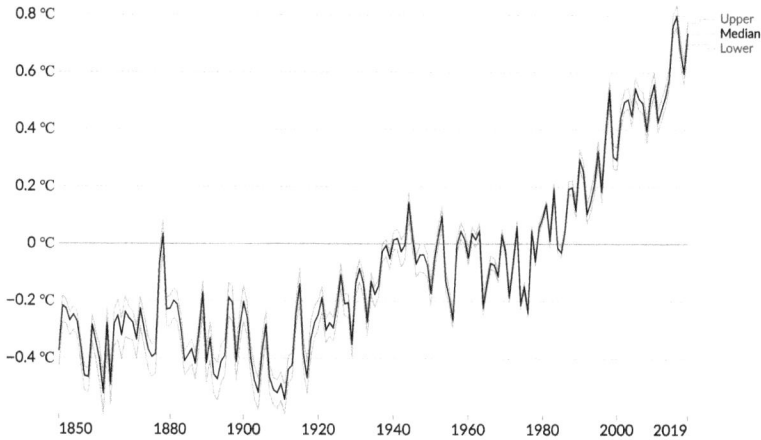

Figure 6.2 *The global average temperature of the Earth up to 2020, compared to the average temperature 1850–1900. The rise over the last few decades is very clear.*

Loss of ice and sea level rise

A direct consequence of global warming, and one of the clearest indications of its impact, has been the loss of the Arctic Sea Ice. Clear evidence of this is available from the NASA National Snow and Ice Data Center satellite, which has monitored the extent of the summer sea ice since 1979. The resulting graph in Fig. 6.3 shows a very clear trend downwards (indicated by the straight line).

In 36 years approximately 2.5 million square kilometres of summer sea ice have been lost. This is equivalent to the area of Scotland every year. If this rate of loss continues, all of the Arctic sea ice will have vanished in 100 years. At the same time land ice has been lost from both Antarctica and the ice sheets on Greenland. We can see this change in Fig. 6.4.

There are a number of consequences of this. The one we most hear about is the loss of habitat for such animals as the polar bears. A second consequence is a change in the salinity of the Atlantic Ocean due to the addition of the fresh water from the melting ice. This could, in the long term, have a direct influence on the ocean circulation patterns, including a shift in the direction of the North

Average Monthly Arctic Sea Ice Extent
August 1979 - 2016

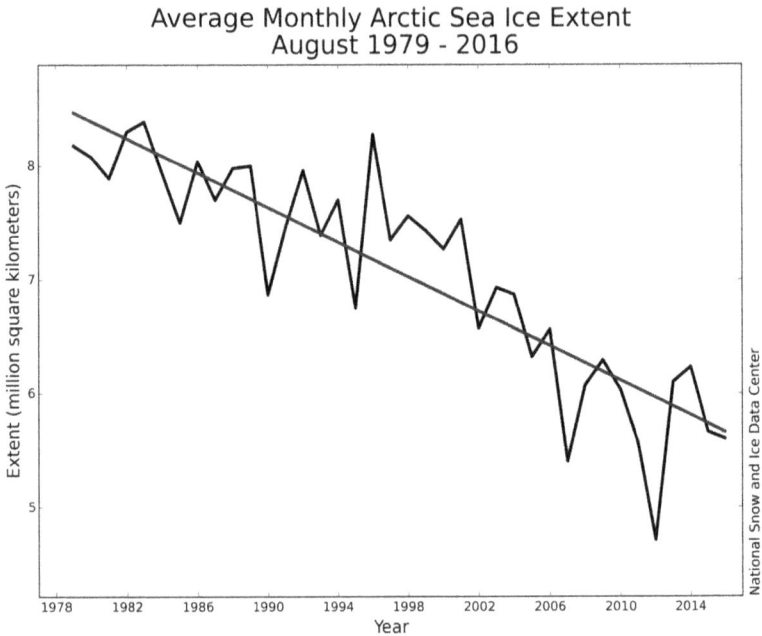

Figure 6.3 *The loss of summer Arctic sea ice as measured by the National Snow and Ice Data Center (NSIDC). (Credit: T. Scambos/National Snow and Ice Data Center)*

Atlantic Drift, which keeps the UK warm. (So ironically global warming at the North Pole could possibly make the UK colder.) A third long-term consequence is that the Earth becomes darker. One of the functions of the ice sheets is that they reflect a large amount of the Sun's energy and keep us cooler as a result. Thus, as the ice sheets retreat so we will warm up. I will return to this topic later when we look at climatic tipping points.

A more immediate, and observable consequence, of the melting ice, and also one which will have significant impact on humanity, is the rise in the average sea level. There are two reasons for this. Firstly the melting of the land (but not the sea) ice adds to the volume of the water in the oceans. Secondly, as the ocean temperature increases due to global warming (as described above), so its volume increases due to the effects of thermal expansion. The increase in sea

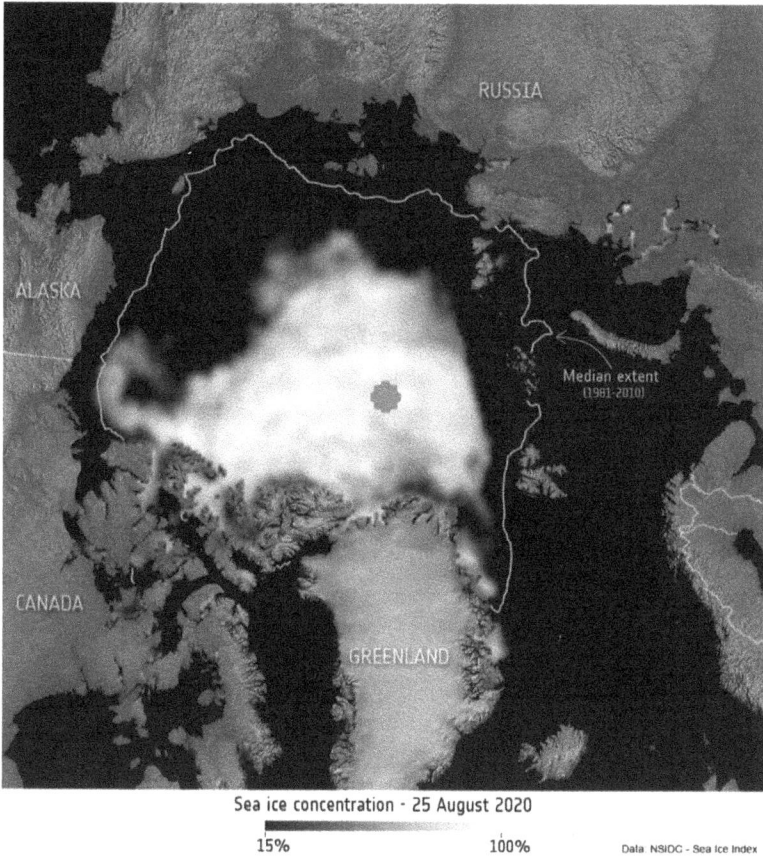

Sea ice concentration - 25 August 2020

15%   100%   Data: NSIDC - Sea Ice Index

Figure 6.4   *The summer Arctic sea ice extent in 2020 compared to the median average of 1981–2010. (Credit: National Snow and Ice Data Center, University of Colorado, Boulder/processed by ESA, https:// www.esa.int/ESA_Multimedia/Images/2020/08/Arctic_sea_ice_concentration_25_August_2020)*

level can be measured using tide gauges and more recently by radar from orbiting satellites. Figure 6.5 shows that the sea level is currently rising at a rate of about 3 mm per year.

The main concern about sea level rise is the impact that it will have on low lying coastal areas, especially when compounded with storm surges and other effects of events such as hurricanes. Sustained sea level rise will make many of

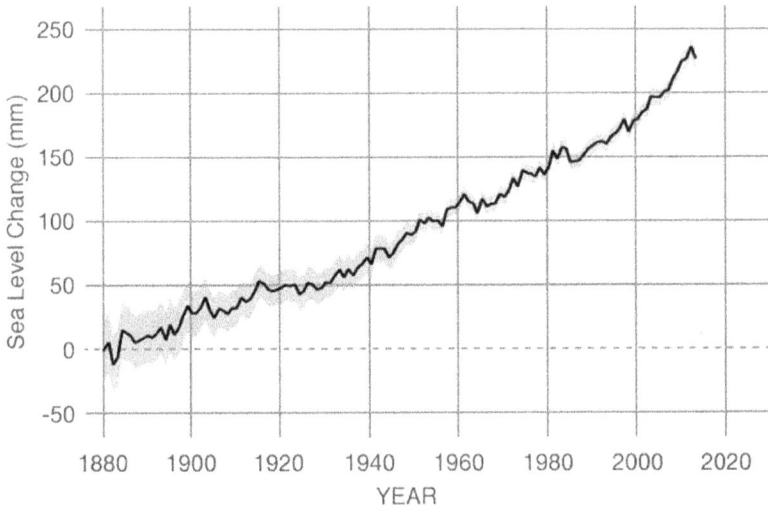

Figure 6.5 *Average global sea level rise, showing a current increase of about 3 mm per year. (Credit: NASA Goddard Space Flight Center)*

the world's cities and coastal regions very vulnerable. Indeed, the 2021 report for the IPCC [23] warns that:

> *Over the next 2000 years, global mean sea level will rise by about 2 to 3 m [7–10 ft] if warming is limited to 1.5°C, 2 to 6 m [7–20 ft] if limited to 2°C, and 19 to 22 m [62–72 feet] with 5°C of warming, and it will continue to rise over subsequent millennia (low confidence).*

and that

> *In some areas, coastal flooding that occurred once a century in the recent past could be a yearly event by 2100.*

Increases in the number of extreme events

You have probably noticed that there have been a lot of big storms recently, for example, the major wind and rain storms in the UK, including the

St Valentine's Day storm in 2014. Other recent extreme events have included the significant 2003 heat wave, which killed thousands of people, the very severe heat wave on the East of North America in 2021, the severe hurricanes affecting the USA since the start of the 21st Century, and extensive flooding in Pakistan. Is this evidence for climate change? The answer is almost certainly YES, but to understand why we need to understand a bit about statistics. We saw above how the average temperature of the Earth has increased by 1 °C since records began. This may not seem very much, after all the daily temperature variation is much more than this. However, this shift in the mean temperature significantly increases the likelihood of extreme temperatures, and related events such as higher rainfall (as a warmer atmosphere can hold more water). To see this, have a look at Fig. 6.6.

The graph on the left shows a typical statistical distribution of temperatures, which is a Bell Curve centred on the mean. The portion on the far right is the tail of the distribution and the area under the tail shows the probability of a hot temperature. The curve on the right shows what happens if the average temperature increases. This has the effect of shifting the whole bell curve to the right. The consequence on the chance of high and extreme weather is profound. The small shift to the right raises the height of the curve in the tail by a very large amount. This in turn dramatically increases the chance of having extreme weather events such as an increased frequency of tropical storms. In the stark words of the 2021 IPCC report [23].

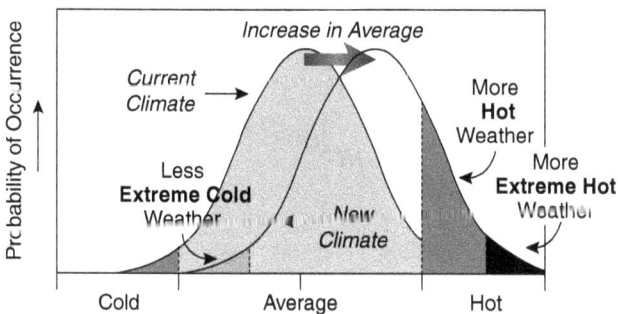

Figure 6.6   *A schematic showing that a small shift in the mean of events (such as the global temperature of the Earth), can dramatically increase the likelihood of an extreme event.*

> *With every additional increment of global warming, changes in extremes continue to become larger.*

Carbon Dioxide increases

It is possible to monitor the amount of Carbon Dioxide in the atmosphere with high precision. For example, the Mauna Loa observatory in Hawaii takes daily measurements of the amount of Carbon Dioxide in parts per million (ppm) in the atmosphere. The disturbing results of these measurements are shown in Fig. 6.7 in what is called a *Keeling Curve*.

The Carbon Dioxide levels vary up and down during the course of a year owing to seasonal changes. However, the overall trend is one of rapid increase. Indeed in the last fifty years the average amount of Carbon Dioxide has risen from 320 ppm to the current value of nearly 420 ppm in 2020. This fact is undeniable, and is mainly due to the burning of fossil fuels such as oil, coal and gas. However, it is the impact of the Carbon Dioxide rise on the Earth's temperature which leads to a lot of controversy and is at the heart of the

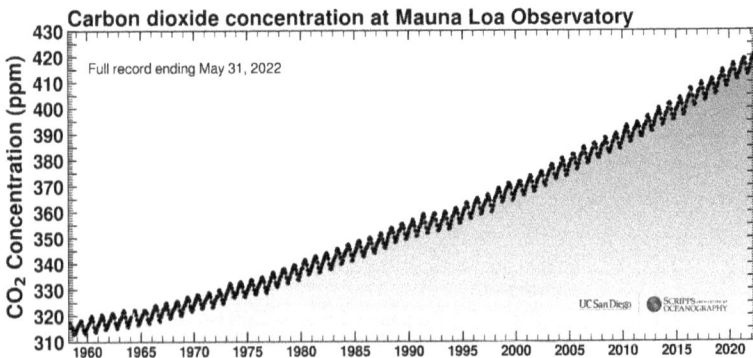

Figure 6.7 *The rise in Carbon Dioxide concentration at the Mauna Loa Observatory, measured in parts per million. The pre-industrial level was 250 ppm, while the current value is nearly 420 ppm. (Credit: Scripps Institution of Oceanography, https://keelingcurve.ucsd.edu/)*

IPCCs recommendations on the need for a 'low Carbon economy'. The key question is whether the increase in Carbon Dioxide levels leads to a temperature increase (due to the 'Greenhouse Effect'), or conversely whether it is a rise in temperature (say, due to natural causes), which has led to an increase in Carbon Dioxide levels. If it is the former then we need to do something to stop a climate disaster. In the latter, we are just the victims of natural climate variations and cannot do anything about it. In practice most of the mathematical models for climate change show that temperature and Carbon Dioxide levels are closely linked with changes on one leading to changes in the other.

For example, a small rise in temperature can lead to a bigger increase in Carbon Dioxide, which in turn leads to an even bigger increase in temperature. We have already seen this effect in the simple model that I described in the Introduction. Later in this chapter I will show how more sophisticated mathematical models of climate change are also consistent with the first viewpoint and not with the second.

A test of some of the most sophisticated climate models used by the IPCC (which will be described shortly) is to use the method of *hind-casting*. In this the climate is set to that at the beginning of the 20th Century and the models are asked to 'predict' the last 100 years of the Earth's climate up to the present day. Two scenarios are typically chosen. In the first the Carbon Dioxide levels are set to the measured values which includes the effect of additional Carbon Dioxide in the atmosphere due to human activity. In the other scenario the Carbon Dioxide levels are kept at the same value that they had in 1900, allowing for natural variations due (mostly) to volcanic activity. The hind-cast tests of the climate models with the observed Carbon Dioxide levels correctly reproduce the variations in the Earth's temperature. This helps to test and validate the models. In contrast the tests of the models with the naturally varying Carbon Dioxide levels, predicted much lower temperatures than observed. As well as testing the models, this experiment strongly demonstrates the role played by human-made Carbon Dioxide in recent global warming. This is illustrated in Fig. 6.8, taken from the 2021 IPCC report [23].

How do we know humans are causing climate change?

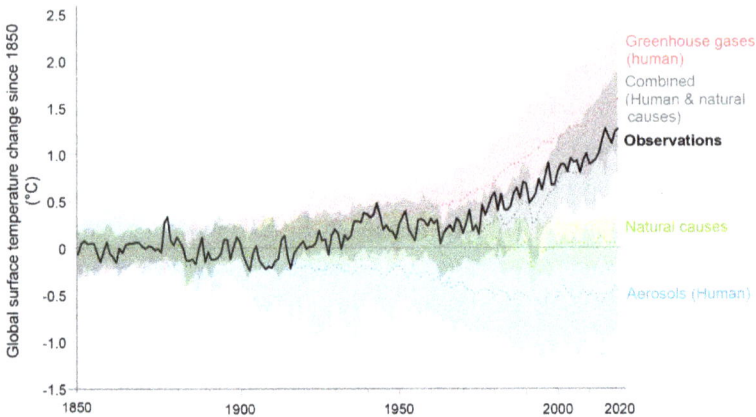

Figure 6.8 *The observed increase in the Earth's temperature over the last 170 years compared with the (grey) model predictions, made using the recorded levels of greenhouse gases, aerosols, and also the natural effects of volcanoes and El Nino. The predictions in grey also include the model uncertainty. There is good agreement with the observations. In the green you can see the same predictions, taking into account the natural causes, but omitting the changes due to greenhouse gases and aerosols. These predictions are well below the observations. This both confirms the models used, and also shows the impact of human activity on the climate. (Credit: IPCC 2021 report [23])*

Past climate changes

A common criticism of the above evidence for climate change is that it is simply natural and not caused by human action. To a certain extent this is correct in that there is no question at all that the climate has changed in the past, and in ways that human beings could not have been responsible for. For example, about 400 years ago we were experiencing a 'Little Ice Age' when temperatures were noticeably cooler than today, and then the Earth gradually warmed up. There is also clear evidence that millions, if not billions of years ago, the Earth has gone through periods of being very cold, and also of being very hot. Reasons for this include changes in the energy levels of the Sun and also the Milankovitch cycles, in which wobbles of the Earth in its orbit change the amount of solar radiation (insolation) reaching the Earth's surface. Whilst these certainly cause changes to the Earth's climate, they are also on long timescales, such as tens of

thousands of years or longer, not on the much shorter timescales of tens of years that we are currently observing.

A further criticism which has been made of the recommendations of the IPCC is that the Earth will naturally regulate its climate as it has done in the past, and that it will do so in the future. However, both of these criticisms fail. In the first case, as we shall see, the observed changes in climate are the opposite of what we would expect from extrapolations of the historical. In the second case, the changes that we are currently seeing are far more rapid than any such changes in the past. It is most unclear whether the Earth's recovery mechanisms can react fast enough to counter the effects of these. I will return to this topic later in this chapter.

Modern climate records taking direct measurements of temperature, rainfall, ice cover, and other climate indicators, really only started when the Met Office was founded in 1850 in the UK. In paleo-climatology, or the study of past climates, scientists use proxy data to reconstruct past climate conditions. Proxy data are preserved physical characteristics of the environment that stand in for direct measurements. Paleo-climate scientists gather proxy data from natural recorders of climate variability such as tree rings, ice cores, bore holes, fossil pollen, ocean sediments, corals and historical data (such as the French grape harvest). By analysing records taken from these and other proxy sources, climate scientists can extend our understanding of climate well beyond the instrumental record. One of the most important of these proxy measurements is the Oxygen-18 isotope. Oxygen occurs in the Earth's atmosphere in various isotope forms. Most of it is the Oxygen-16 isotope, but a smaller amount is the heavier Oxygen-18 isotope. The ratio between the Oxygen-16 and Oxygen-18 water molecules in an ice core, helps to determine past temperatures and snow cover. The heavier Oxygen-18 isotope condenses more readily as temperatures decrease and falls more easily as rain or snow, while the lighter Oxygen-16 isotope needs colder conditions to precipitate.

The temperature over the last 2000 years as estimated by various proxies is illustrated in Fig. 6.9. Different proxies lead to slightly different reconstructions,

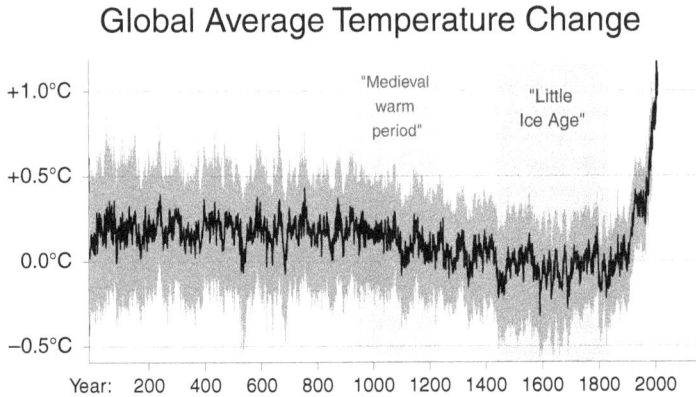

Figure 6.9 *The change in the Earth's temperature over the last 2000 years as estimated from different proxy measurements. Note the 'hockey stick' rise in the temperature in the last 200 years. (Credit: RCraig09, https://commons.wikimedia.org/wiki/File:2000%2B_year_global_temperature_including_Medieval_Warm_Period_and_Little_Ice_Age_-_Ed_Hawkins.svg)*

but the overall pattern is clear. There was a gradual warming over the first 1000 years to give the Medieval Warm period, during which Europe was warm although the rest of the Earth was rather cooler. We then see a gradual cooling resulting in the Little Ice Age which was probably due to a reduction in solar activity. The Little Ice Age is then followed by a very rapid warming as we approach the present day. Indeed, the global temperature has risen more since 1970 than in any half century going back over 2000 years. To arrive at a period warmer than 1850–2020, we have to go back to before the last Ice Age, more than 100,000 years ago. The very rapid recent rise is often called the 'hockey stick'. It has been repeatedly reconfirmed by many different climate researchers.

By drilling deep into the Antarctic ice, and measuring the Oxygen we can find out the temperatures of the past million or so years. The results are presented in Fig. 6.10, and are remarkable. In this famous figure the top two graphs show the temperature over the last 450 thousand years, as estimated from two different ice core samples called European Projects for Ice Coring in Antarctica (EPICA) and Vostok. The lowest graph shows the estimated volume of the ice covering the Earth over the same time period. This figure shows clear evidence for four complete ice ages. In a typical glacial cycle the Earth gradually cools and the

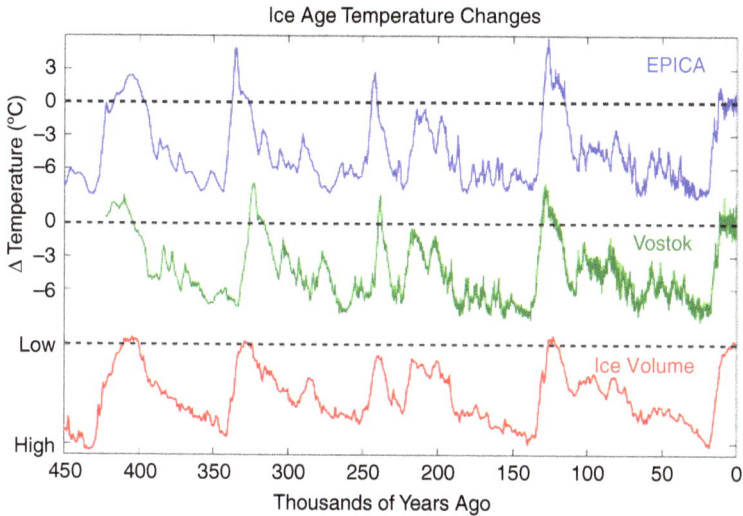

Figure 6.10 *The temperature and ice cover of the Earth over the last 500,000 years. The top two graphs show the temperature derived from two different ice core measurements. The lower curve is the total ice cover. In this figure you can see evidence for four complete glacial cycles with the Earth's temperature dropping in the ice ages. The most recent ice age ended around 10,000 years ago. (Credit: https://commons.wikimedia.org/wiki/File:Ice_Age_Temperature.png)*

ice volume increases. The ice age then ends in a rapid warming period. The most recent such period was the end of the younger Dryas Ice Age about 10,000 years ago. This of course led to the modern age, in particular the start of agriculture and the growth of civilisation. What I find remarkable about this graph is its regular periodic behaviour. In particular we see ice ages, with large variations in temperature appearing, almost by clockwork, every 100,000 years. What is also interesting is that as we go further back in time this cycle changes and is replaced by much smaller changes in temperature every 40,000 years. We will have a look at what causes the ice ages later in this lecture. But one prediction from this regular graph is very clear. If nature was left to itself then we should be entering the cooling phase of the next ice age. Instead, things are getting warmer.

A plot of the historical values of Carbon Dioxide, given in Fig. 6.11, is equally striking. Again we can see natural cycles in the Carbon Dioxide levels between 180 ppm and 300 ppm, which are in synchrony with the temperature and ice

CARBON DIOXIDE OVER 800,000 YEARS

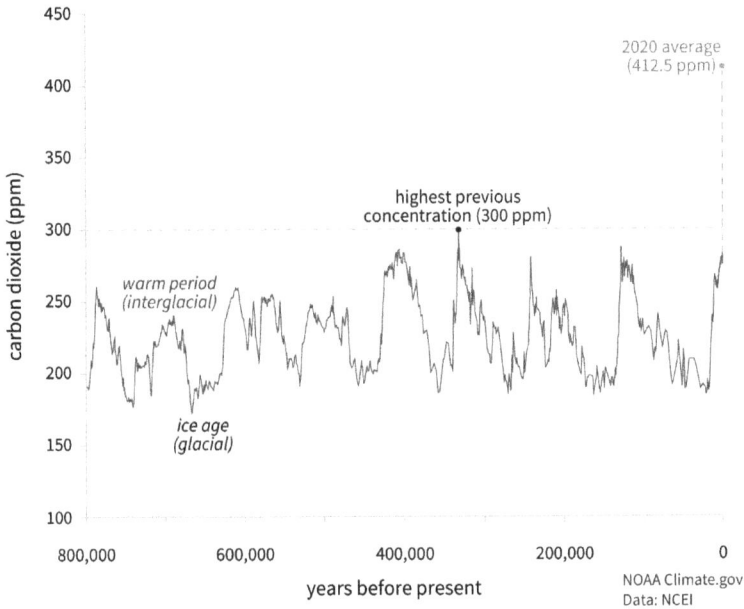

Figure 6.11 The historical variation in Carbon Dioxide levels. These have in the past always been below a value of around 270 ppm. However, in recent years we have seen an unprecedented rise. (Credit: NOAA Climate.gov/NCEI, https://www.climate.gov/media/12987)

cover variations. Yet as we approach the present day, and in particular since the start of the industrial revolution, the Carbon Dioxide levels have increased extremely rapidly to the current value of about 420 ppm. This human-made rise is far more rapid than any natural variation. The 'hockey stick' in the rise in Carbon Dioxide is closely linked to the same sudden rise in temperature that we saw earlier.

3. How we model climate change

How climate models are derived

As I said in Chapter 5 it is hard to predict anything, especially in the future. Climate change is no exception to this. There are various reasons for

this difficulty. The climate is very complex, it is hard to obtain good data (especially of the initial states), the equations for climate are hard to solve and may have multiple solutions, chaotic behaviour is always present, and it can be hard to distinguish natural effects from human intervention. A careful mathematical approach, based on the ideas in Chapter 1, is therefore essential if we are to construct climate models with any degree of reliability, and also to be able to assess the level of uncertainty in the predictions of that model.

Climate change models are complex and large. If we link them to models of the effect of climate change on the economy and society then they are even more complex! Such models have millions, if not billions, of lines of computer code in them. So, how are the models constructed, tested and do we believe them?

As in many of the mathematical models that we considered in the last two chapters.

> *All climate models start from the laws of physics.*

These laws have been carefully tested and validated over centuries. Most climate models also are based on weather prediction codes, which are tested every day! Basically the climate that we see arises from the interaction of the energy coming from the Sun with the atmosphere, the oceans, the ice and the vegetation.

Weather forecasting

The basic laws of physics for *weather forecasting* that we first met in Chapter 1, are the Navier–Stokes partial differential equations of fluid motion on a rotating sphere (which describe the evolution of the momentum and energy of the air and the oceans). These are coupled to the laws of thermodynamics (which describe the evolution of the temperature, and the effect of heat from the Sun on air, water and water vapour). Together these equations are:

$$\frac{Du}{Dt} + 2f \times u + \frac{1}{\rho} \nabla P + g = v\nabla^2 u,$$

$$\frac{\partial \rho}{\partial t} + \nabla \cdot (\rho u) = 0,$$

$$C\frac{DT}{Dt} - \frac{RT}{\rho}\frac{D\rho}{Dt} = \kappa_h \nabla^2 T + S_h + LP,$$

$$\frac{Dq}{Dt} = \kappa_q \nabla^2 q + S_q - P,$$

$$p = \rho RT.$$

Again, a very scary looking set of equations, but very useful as these are precisely what weather forecasters use every day. Whilst I do not expect you to be able to follow the mathematical details of each equation, I can try to explain what each equation means. In these equations, u is the velocity of the air and the first equation is Newton's law of momentum, balancing forces and pressures with accelerations. The second equation expresses the fact that the total amount of the air of density ρ is conserved. The third equation is one of the equations of thermodynamics and describes how the temperature T of the air evolves. The fourth equation shows how the moisture q of the air changes. The final equation is the universal *equation of state of the atmosphere* which relates its pressure P to its density and temperature.

These equations for the weather are solved *every six hours* by the super-computers at the Met Office, to produce a five-day weather forecast, and the predictions from these equations are tested every day. They are, however, much too complicated to solve by hand. Instead we solve them by using a computer. The first idea of doing this came from the British all round genius Lewis Fry Richardson in 1917, and his envisaged computer was a room full of students. His idea was to divide up the Earth, and its atmosphere and oceans, into a large number of small cubes, or tetrahedrons. The equations of the weather are then approximated over each cube in a procedure called discretisation which is illustrated in Fig. 6.12.

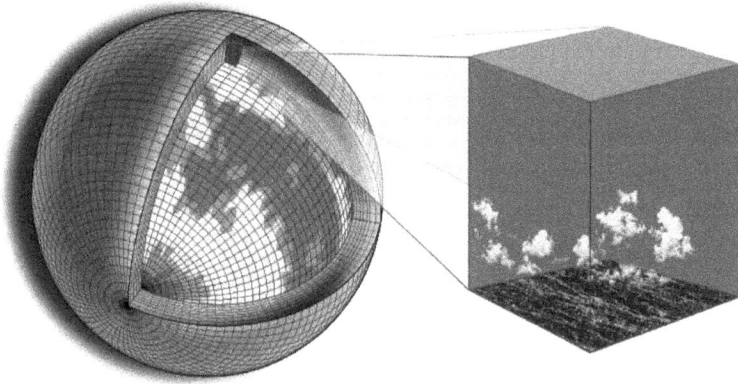

Figure 6.12   *The process of discretisation. The Earth's atmosphere (and oceans) are divided up into a large number of small tetrahedrons, and the weathers calculated in each cube. The results are then combined to give a complete weather forecast.*

Exactly the same idea is still used to produce the weather forecast today. The smaller the cubes used, the more accurate the forecast, but the more computer power which has to be used to do the resulting calculations. For the UK forecast each such cube has a size of about 1.5 km on each side. In a typical forecast there are about a billion such discrete equations. To solve them involves inverting very large matrices and such a calculation takes about one hour on a super computer. To see more I recommend the WXCHARTS website [24].

Climate forecasting

To see how the climate evolves we have to augment these equations. This process is hard for a number of reasons. *Firstly*, to predict climate changes we have to forecast many years ahead (sometimes thousands or millions of years). *Secondly*, we have to include a lot of extra physics, chemistry and even biology. This includes ocean currents, sea ice, land ice, solar physics, complex atmospheric chemistry, vegetation (on both land and in the ocean), animals, clouds, permafrost, and greenhouse gases. *Thirdly*, the systems we are looking at in climate modelling are very nonlinear and may well have chaotic solutions, such as I described in Chapter 5. *Finally* (and most uncertainly) we have to account for the current, and future impact of humanity, including such effects as the changes amount of Carbon Dioxide in the atmosphere as the result of the burning of fossil fuels, changes in farming practices or of cutting down the rain forests. Because of the size and complexity of the resulting systems it is hard to check them, change them and to run them. It is also hard to interpret the results as they produce a mass of data (a billion data points for each forecast) which is hard to analyse and even harder to store.

The science-fiction author Robert Heinlein (author of *Starship Troopers* amongst other books) made the following quote (sometimes attributed to Mark Twain) in his book *Time Enough for Love* in 1973.

> *Climate is what you expect, and weather is what you actually get.*

I completely agree. What this quote means is that when we say 'climate' we mean an expression of the range of possibilities of the state of the Earth's atmosphere and oceans (the probability distribution), and then weather is a sample from this. To predict the climate we need to assess what this range of possibilities is, and also how it changes with time. This is a different (although closely related) task to that of finding the day-to-day changes in the weather.

The sort of models we then use

Climate models have been evolving considerably over the last decades, both in accuracy and in complexity. They evolve in parallel both with faster computers and also constantly improving mathematical models [25]. Figure 6.13 shows how the climate models have developed in complexity over the last 40 years.

However, in order to use and to run a climate model, certain approximations have to be made in comparison to a weather model. This is needed because, as we have seen, climate models are much more complex than weather models. Furthermore, instead of looking five days into the future, a climate model must look decades, or even centuries, into the future. For this to be possible the spatial resolution of a climate model is much coarser than in a weather forecast. For example the cubes for the discretisation may be 100 km wide or more.

The Development of Climate Models: Past, Present and Future

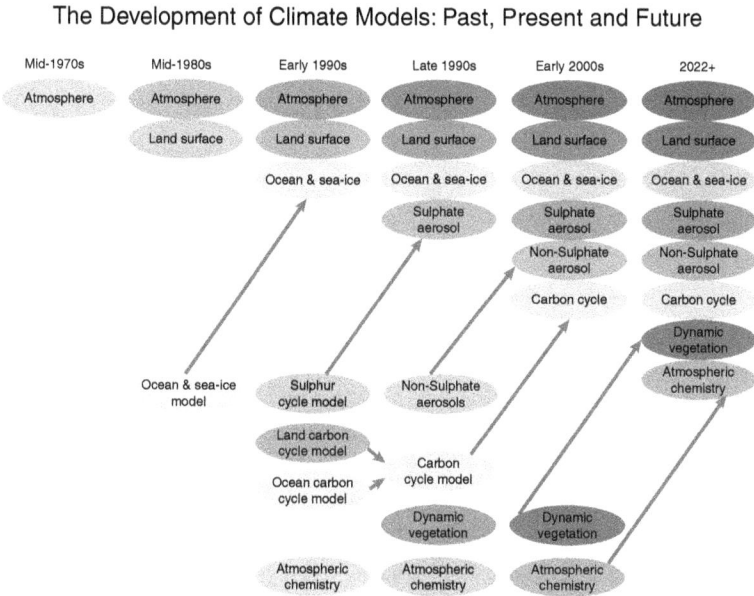

Mid-1970s	Mid-1980s	Early 1990s	Late 1990s	Early 2000s	2022+
Atmosphere	Atmosphere	Atmosphere	Atmosphere	Atmosphere	Atmosphere
	Land surface	Land surface	Land surface	Land surface	Land surface
		Ocean & sea-ice	Ocean & sea-ice	Ocean & sea-ice	Ocean & sea-ice
			Sulphate aerosol	Sulphate aerosol	Sulphate aerosol
				Non-Sulphate aerosol	Non-Sulphate aerosol
				Carbon cycle	Carbon cycle

Figure 6.13 *The inclusion of successive levels of complexity into the climate models used by the Hadley Centre in the UK. The arrows show when a model was developed and then included into an operational climate forecast.*

Also the time steps are longer, and often the models look to find averaged quantities. There is a trade off in the amount of spatial resolution against the length of time that these models can predict into the future. The different codes used by the Met Office and Hadley Centre are illustrated in Fig. 6.14.

In this image we can see the difference in application between a local weather forecasting code such as UK4 and all of the models in the bottom left part of the image, and a general circulation model (GCM) such as HADGEM3 in the top right of the image.

How climate models are tested

A *General Circulation Model* (*GCM*) is a very complex piece of software with many millions of lines of code. Errors can arise in the way that the physics is represented, the algorithms used to solve that physics, the coding up of those

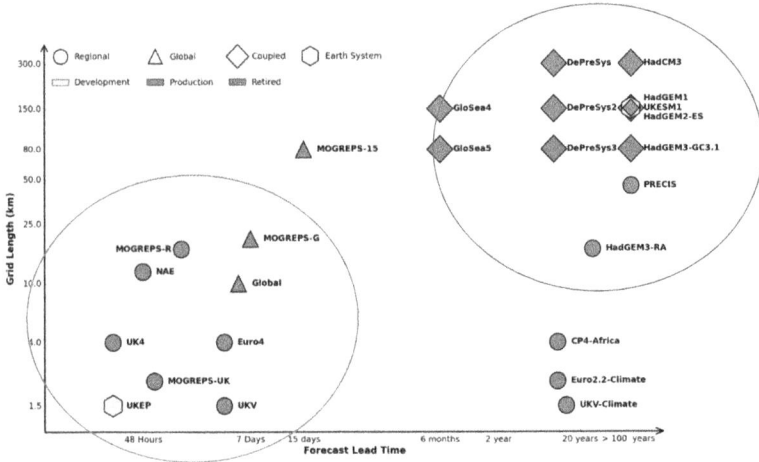

Figure 6.14 *A comparison between the weather models (bottom left) and the climate models (top right) used by the Hadley Centre. In this figure the x-axis shows how far into the future predictions are made, and the y-axis shows the spatial resolution of the forecast. (Credit: With kind permission from the UK Met Office)*

algorithms, the data that is fed in to the calculation and the initial conditions used to start the whole system off. Because the climate models are based on weather models, one aspect of the testing is always available. Indeed, weather models are tested by comparing them against reality every six hours. A modern weather forecast updates its prediction constantly by comparing its predictions against data in a process called *data assimilation* [25]. Any systematic error in the code would quickly reveal itself in this process. A second check is that through mathematical arguments we can check the convergence of the algorithms used through the methods of *numerical analysis*. Thirdly, all modern climate algorithms assume that they are working with uncertain data. By using techniques from probability and statistics it is now possible to quantify this uncertainty so that we have a reasonable idea of how accurate our forecasts may be. In particular we can build in uncertainty due to the model, the initial conditions and the 'scenario' (estimates of future human activity, for example). A further way of testing a particular climate model is to compare it with the predictions of different models. There are many different climate centres worldwide. These use models which differ in the way they approximate the physics, the way they solve the resulting partial differential equations (for

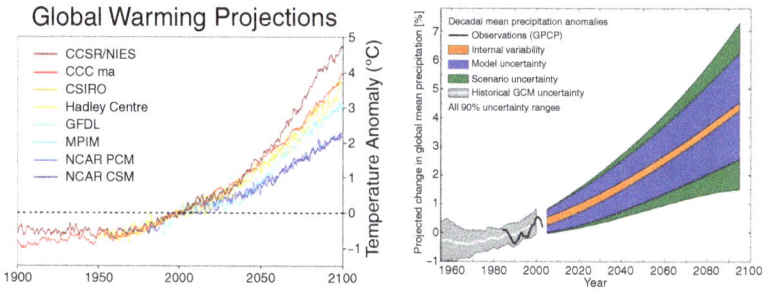

Figure 6.15 *The left figure shows the projected increase in the average temperature of the Earth till 2100 from a number of different climate centres. The figure on the right is a prediction of the rising temperatures from the Hadley Centre in the UK, together with various forms of uncertainty. (Credit: With thanks to Ed Hawkins)*

example, by using a finite volume or a spectral method), and the various assumptions of human activity. Such centres include the NCAR in the USA, Hadley Centre in the UK, CSIRO in Australia, and CCSR in Japan (see Fig. 6.15). On the right we see the Hadley Centre predictions for the UK temperature with associated uncertainty, and on the left the predictions of global temperature using these various models (with the US predicting the lowest, the Japanese the highest and the UK in the middle).

Unlike a weather forecast which is tested (for better or worse) every day, it is impossible to test a climate forecast directly against future data unless we are prepared to wait decades for the result. Instead we test them against past data using the hind-casting method we looked at earlier. These are run over the 150 years for which we have reliable climate data and provide a careful check of all of the predictions of the model with what is observed. Essentially if a climate model can predict the past, then we have a good cause to believe that it can predict the future. As models for climate have now been around for long enough for certain climate variations to have taken place, we can also compare their predictions with variations of temperature and sea level rise observed over the last twenty or so years. Again the predictions have been good (and have been carefully monitored by the IPCC as described in [20]). Nevertheless they have been faced with the very real problem of estimating how much Carbon Dioxide would actually be released into the atmosphere due to human activity. As with all human based issues, this is the hardest thing for any mathematical model to predict with any level of accuracy.

4. Some 'simple' mathematical models of climate

There are several problems with using the GCM climate models preferred by the IPCC. The first is that their sheer size makes them very expensive to run, and the computers running them use a lot of energy in the process. For example, the Met Office Cray XC40 supercomputer makes 14,000 trillion arithmetic operations per second, and uses 2.7 megawatts of power when it is running. (We will see how this electricity is generated in the next chapter.) One of the ironies of modern climate science is that the computers doing the calculations contribute, by doing them, to global warming. They also take a long time to run, so that it can take days to get a forecast of the climate in 100 years. To get an estimate of the climate in a million years is quite impossible with a GCM. Thirdly, it is hard to do 'what if?' experiments, in which we look at the effect of changing different parts of the model (such as the amount of greenhouse gases). Finally (and crucially for someone like me) the models are much too complex to be analysed by hand, so whilst we may be able to predict things, it is very hard to explain why we are seeing what we see. We therefore work with a hierarchy of different models of increasing complexity as seen in Fig. 6.16. At the simplest level are energy balance models (EBMs), an example of which we met in the introduction. At the next level up are Box Models which divide the Earth into a number of large boxes. Next come Earth Intermediate Complexity models (EMICS) or Reduced Climate Models (RCMs), then there are the (atmosphere and ocean) General Circulation models (GCMs) described above — these in turn are part of very complex Earth System Models (ESMs) also described in the Introduction.

Energy Balance Models (EBMs)

The simplest of all of these models (and yet still with a good degree of predictability as we saw in the introduction) are the Energy Balance Models

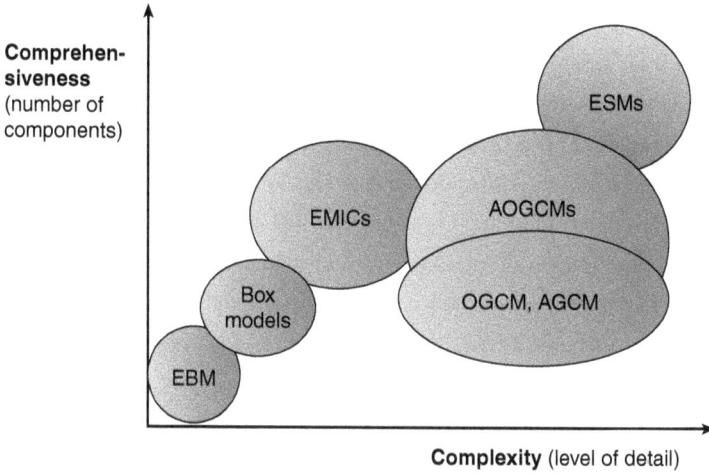

Figure 6.16 *A comparison of the different types of climate model, from the simplest Energy Balance Models (EBMs) which have limited complexity, to the highly complex Earth System Models (ESMs). In this figure A(O)GCMs are atmosphere (and ocean) general circulation models.*

(EBMs). Whilst this model is really simple (indeed it can be implemented in Excel), it can be used to inform us on the implementation of the 2016 Paris Agreement to keep the Earth's temperature to below 1.5°C, and even as we advertised earlier, to find the mean temperature of the Moon.

We will now construct an EBM using the steps outlined in Chapter 1, and by doing so we will justify the model described in the introduction to this chapter.

Step 1 is to observe that all of the energy coming to the Earth from the Sun must either be radiated back into space or used to heat up the Earth and its atmosphere. Balancing these two (as energy must be conserved) gives us the basis of a simple climate model.

In **Step 2** we consider the basic physical laws. The whole Earth is being heated up by the Sun and releasing its energy back into space. We take $S(t)$ to be the radiation from the Sun, which at any one time illuminates one half of the Earth's surface. This illumination is primarily seen as short wavelength radiation. A proportion of this radiation is reflected back by the Earth, where a is the Albedo. If $a = 1$ then all of the radiation is reflected, and if $a = 0$ then none of it is. So the total radiation reaching the Earth's surface is $(1 - a)$ times the average solar radiation. This radiation heats up the Earth and its atmosphere, which in turn re-radiate it back into space as long wavelength radiation (infra-red). The amount of this radiation is given by the black body radiation law, in which the radiation is proportional to the fourth power of the absolute temperature. Not all of the radiation from the Earth goes back into space. Indeed, a significant proportion of it is absorbed in the atmosphere and reflected back to Earth. This absorption is the result of the greenhouse gases in the atmosphere, such as Carbon Dioxide, Methane and water vapour.

For **Step 3** we look at how to simplify this system. A major simplification is that we can model these processes by making the approximation that the Earth has an average temperature of T_E and that the atmosphere has an average temperature of T_A. (This is of course a huge simplification, as it ignores the change in the temperature from one part of the Earth to another. However, it is still useful, and gives a good estimate for T_E.)

For **Steps 4** and **5** we now formulate these processes as (algebraic) equations, and then we solve the equations.

We start by looking at the radiation $S(t)$ coming from the Sun. If the Earth has radius R then its surface area is $4\pi R^2$. The area of the disc that the Sun illuminates is in contrast πR^2 and the flux of the solar radiation is $\pi R^2 S(t)$. The *average* solar radiation on the Earth's surface is then $\pi R^2 S(t)/4\pi R^2 = S(t)/4$. This value can be measured and

$$S(t)/4 = 342 \text{ W/m}^2.$$

The amount of short-wave solar radiation which is not reflected back into space passes through the atmosphere. The amount of this which reaches the Earth, warming it up is given by

$$\tau_{SW}(1-a)S(t)/4,$$

where τ_{SW} is the short-wave transparency of the atmosphere. The atmosphere is mostly transparent to shortwave radiation and we have $\tau_{SW} = 0.9$. The remainder of the Sun's energy

$$(1 - \tau_{SW})(1 - a)S(t)/4$$

helps to heat up the atmosphere.

The Earth heats up to a temperature T_E and radiates an energy flux F_E. This can be calculated by using the black body radiation equation and we have

$$F_E = \sigma T_E^4,$$

where $\sigma = 5.67037 \times 10^{-8}$ Wm^{-2}K^{-4} is the *Stefan–Boltzmann constant*.

This flux is in long-wave infra-red radiation. As it passes through the atmosphere most of it is absorbed and this also helps to heat the atmosphere up. Now we introduce another factor τ_{LW} which is the *transparency of the atmosphere to long-wave radiation* in the infra-red. Most of the infra-red radiation is absorbed due in part to the action of the greenhouse gases. As a result $\tau_{LW} = 0.2$ which is a lot smaller than the short-wave transparency. (In other words the atmosphere lets light in but is reluctant to let heat out.) The amount of the Earth's radiation which goes into space is then $\tau_{LW} F_E$. The remainder of the radiation from the Earth is

$$(1 - \tau_{LW}) F_E,$$

which also helps to heat up the atmosphere. When this is combined with the energy $(1 - \tau_{SW})(1 - a)S(t)/4$ from the Sun, it raises the temperature of the atmosphere to a temperature T_A. The atmosphere itself then radiates energy, part of which is radiated into space and the other back to Earth. The energy flux from the atmosphere divides into F_A towards the Earth and F_A into space. The value of F_A is also given by the black body formula

$$F_A = \sigma T_A^4.$$

As this radiant energy is also in the infra-red, a proportion $\tau_{LW} F_A$ reaches the Earth's surface and helps to heat it up.

Now we can do some balancing of the energy fluxes. On the Earth's surface we have a balance of the incoming long-wave heating from the atmosphere and the short-wave heating from the Sun, with the energy flux from the Earth so that

$$F_E = \tau_{SW}(1-a)S(t)/4 + \tau_{LW}F_A.$$

Above the atmosphere there is a balance between the energy coming in from the Sun and the energy leaving from the Earth and from the atmosphere gives:

$$\tau_{LW}F_E + \tau_{LW}F_A = (1-a)S(t)/4.$$

Adding these two energy balance equations together, cancelling the term $\tau_{LW}F_A$, which appears on both sides of the formula, and rearranging we have

$$\frac{(1-a)S(t)}{4} = \left(\frac{1+\tau_{LW}}{1+\tau_{SW}}\right)F_E = \left(\frac{1+\tau_{LW}}{1+\tau_{SW}}\right)\sigma T_E^4.$$

It is convenient to set

$$e = \left(\frac{1+\tau_{LW}}{1+\tau_{SW}}\right).$$

Here e is the *emissivity of the atmosphere* which is a *proportion of the radiated energy from the Earth which goes back into space*. The value of e is 1 when there is no atmosphere, such as on the Moon, and is currently measured to be about 0.605 on the Earth. (This is slightly different from the value given in the above formula due to some other energy transfer mechanisms which I have not included in the above model, such as evaporation and thermal convection.) It then follows that

$$\frac{(1-a)S(t)}{4} = e\sigma T_E^4.$$

Rearranging this formula allows us to calculate T_E from the equation:

$$T_E = \left(\frac{(1-a)S(t)}{4e\sigma}\right)^{1/4}.$$

We call this the *energy budget equation,* and it is the main prediction from our model. It is of course, exactly the model that we considered in the introduction.

For **Step 6** we need to make a check of the energy budget equation against some data. In fact we have already done this in the introduction where we calculated the average temperature of the Earth to be

$$T_E = 288 \text{ K}.$$

This value is about right, and shows that the model is along the right lines.

As a separate test, and as advertised, we can work out the temperature of the Moon. On the Moon we have $e = 1$ as it has virtually no atmosphere (due to its reduced gravity) and all of the infra-red energy escapes back into space. For the same values of the other constants we can predict a mean temperature $T_M = 254$ K. This is again about right, although, due to its slow rotation, the temperature of the Moon is very different on the side which faces the Sun from that which is in darkness.

Step 7 asks us to consider improvements to this simple model. This takes us deeper into the subject of climate modelling, and we will do this shortly. However, the Energy Balance Model whilst simple, is robust enough to allow us to make some predictions.

Step 8 Indeed, having made the checks, we can use the energy budget formula to predict possible future climate change and to influence policymakers.

Key to these predictions is the transparency of the atmosphere to long-wave radiation τ_{LW}. The lower the value of the transparency τ_{LW} the more that infra-red radiation is absorbed. In particular as τ_{LW} decreases then so does the emissivity e. As we saw in the introduction, it follows directly from the energy budget formula that if the emissivity e *decreases* then the mean temperature of the Earth T_E *increases*. It is this prediction which gives us a direct link between Carbon Dioxide levels and the Earth's temperature. It is a scientific fact that the long-wave transparency of the atmosphere depends directly on the composition of the *greenhouse gases* in it. (These gases are called greenhouse gases because they

Table 6.1 *The change in the emissivity due to the impact of Carbon Dioxide.*

Level of Carbon Dioxide (ppm)	Emissivity e_{CO_2}
200	0.194
400	0.14
600	0.108
800	0.085

act rather like the glass in a greenhouse to reflect heat back to the Earth.) Carbon Dioxide is one of these and the level of Carbon Dioxide contributes to the emissivity (alongside that due to the other greenhouse gases of water and Methane). The more Carbon Dioxide there is in the atmosphere, the lower its transparency to long-wave radiation, and hence its emissivity. A table of the calculated emissivity due to Carbon Dioxide alone is given in Table 6.1. The conclusions from this table are unambiguous. As the level of Carbon Dioxide *increases* so the emissivity *decreases*. Hence, the temperature T_E increases. *Thus the huge increases in the levels of Carbon Dioxide in recent years are, according to this model, a clear contribution to global warming.* This we can tell all policymakers.

It is worth saying that similar predictions on the increase in the mean temperature of the Earth can be made when there is an increase in Methane levels. These increase both due to modern agriculture (in particular farm animals) and the melting of the Arctic permafrost. (Whilst water moisture is also a greenhouse gas, it is less important to temperature change as it leaves the atmosphere as rain or snow, whereas Carbon Dioxide and Methane stay in the atmosphere for a long time.)

Using the model we can start to consider the issues related to the Paris Agreement and its impact on policymakers. In particular a value of 1.5 K has been agreed by the Paris Agreement [21] and felt to be the maximum sustainable value of the Earth's temperature rise over the next 100 years.

The **Equilibrium Climate Sensitivity (ECS)** is the global mean surface warming which follows after a *doubling* of the current levels atmospheric

Carbon Dioxide. If $W = S(1 - a)/4$ is the total amount of energy reaching the Earth, then it follows from the energy budget formula that

$$dW/dT_E = 4e\sigma T_E^3.$$

If we use the above values for e and σ and set $T_E = 288$ K, then

$$dW/dT_E = 3.25.$$

It is estimated that doubling atmospheric Carbon Dioxide has the effect of changing W by about 4 W per metre squared. From the above this would indicate a rise in temperature of 1.23 K. This is below 1.5 K. This might lead to grounds for optimism that we can sustain a doubling of the amount of Carbon Dioxide in the Earth's atmosphere and still stay safe. However, the climate is more sensitive than this simple model implies because there are other feedback mechanisms in the climate we have not considered in the simple model. A significant one of these is the effects of ice melting, which we will look at presently.

Other climate models, and the prediction of the ice ages

As we said above, Step 7 is a window into a vast range of other (reduced) climate models. All of these make many simplifying assumptions depending upon the amount of detail that they need to predict and the time-scale over which they have to be applied. Mathematical models exist, for example which make a fair prediction of the roughly four yearly El Nino warming of the Southern Pacific Ocean. Similarly, there are mathematical models which aim to predict what will happen to the currents in the Atlantic Ocean if the Arctic ice all melts.

One of the most intriguing questions in climate change is what causes the ice ages. As we have seen the ice ages occur roughly once every 100,000 years and predicting them is important in helping us to understand what the climate will do in the next 1000 years. Roughly what has happened in the 'recent' series of six ice ages is that the Earth has cooled down over a period of around 100,000 years, during which it has had extensive glaciation (the glacial period). It then suddenly warms up in a few thousand years (the inter glacial period). It then cools down and the cycle repeats. Despite many years of trying we are still very far from understanding both what causes the ice ages, and also why they have such a regular period. A popular theory for many years [26] was that they are simply caused by changes in the Earth's orbit. As the Earth goes round the Sun it spins like a top. As it does so, then over many thousands of years the angle of the axis of this spinning top changes. This alters the amount of incoming solar radiation received by the Earth in a manner which can be calculated. As a consequence, we know that the incoming solar radiation varies up and down with a period of roughly 40,000 years. These changes are called the *Milankovitch Cycles*. However, this theory neither predicts the period correctly or explains why about 750,000 years ago the frequency of the ice ages changed from 40,000 years to 100,000 years, in the so-called Mid Pleistocene Transition (MPT). Predicting the ice ages over such a long period is well beyond the capabilities of a GCM and hence they must be studied using a simpler model such as a box model. Indeed, predicting the ice ages requires a model which is more sophisticated than an energy balance model and less sophisticated than a GCM. Doing this illustrates one of the main hazards of climate modelling. There are many possible simplifications of the full climate model, and there is far from a consensus as to which one should be used to predict the ice ages. In my own research I have counted over 30 so far. A model that I am currently studying, together with my team of PhD students, was proposed in 2004 by Paillard and Parrenin (and called the PP04 model) [30]. This model looks at the close coupling of temperature, atmospheric Carbon Dioxide and overall ice levels. This model assumes that the Earth is heated quasi periodically by the Sun due to the Milankovitch cycles above. During a glacial period the glaciers slowly advance and the Earth cools down. At the same time the atmosphere cools and cannot retain its Carbon Dioxide and this is stored deep down in the oceans. However, the oceans can only store a certain amount of Carbon Dioxide and it is released when it gets to a critical value and the oceans lose the ability to store it any more. When the Carbon Dioxide is released from the oceans into the atmosphere there is a sudden warming up of the Earth due to the Greenhouse Effect that we looked at in the last section. When all of the Carbon Dioxide has been released the Earth enters another glacial cycle. This model both predicts the right period of the ice ages and possibility of the MPT — so I am hopeful

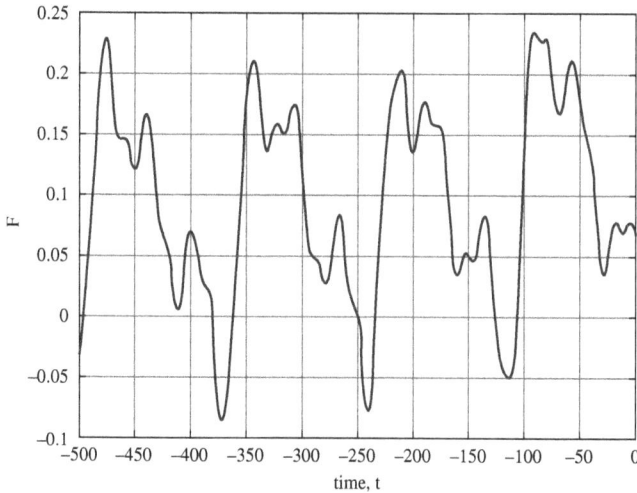

Figure 6.17   *The predictions of the Earth's climate over a 500,000 year period made using the PP04 model. Here F is a climate proxy related to the amount of Carbon Dioxide in the oceans. The predictions from the model are qualitatively similar to the data for the ice ages, shown in Fig. 6.10.*

that we are along the right lines. In Fig. 6.17, we can see is the prediction of a measure F of the climate state as a function of time (expressed in thousands of years) which is given by this model in [30]. You can compare this prediction to the ice age temperature and ice cover values shown in Section 2.

5. Common criticisms of climate models

I do a lot of work on climate models and give many talks about them to many different audiences. This leads to a vibrant correspondence in which I see many questions raised about climate models and their applications and predictions of rapid change due the actions of humanity. Common amongst these are the following:

1. Surely the most important effect on the climate is the Sun and you have missed this out of your models.
2. The climate has always been changing. What we are seeing now is just what you would expect from natural variation.

3. Chaos theory tells us that we can't predict anything that far into the future.
4. Carbon Dioxide rise always comes after temperature rise.
5. Climate models leave out important factors such as volcanic action.
6. Climate models never make accurate predictions.
7. Unprintable, ad-hominem attacks on climate scientists, who are (apparently) only in it for the money, or for political reasons.

In answer to these:

1. The variations in the radiation from the Sun are certainly included in climate models (you can see their effect in the term $S(t)$ in the energy budget equation), and these variations have a significant effect over the long term (for example they were one of the main causes of the Little Ice Age). At the moment the Sun is slowly decreasing its output so if the Sun was the main reason for climate change then we should be seeing the Earth cooling down, rather than warming up.

2. Yes, the climate does change naturally but (as we have seen) on a much longer timescale than we have been seeing in recent years. Natural variations (due for example to the Milankovitch cycles or variations in the output of the Sun) occur over time periods of the order of thousands of years or longer. The rapid changes in temperature that we have been seeing recently have occurred over at most tens of years. This is far too fast to be due to natural variation, and must instead be due to human activity.

3. This is exactly what we looked at in Chapter 5. It is the average properties of climate that are important, and models with chaotic solutions have very predictable average behaviour.

4. In the past Carbon Dioxide and temperature values have very closely coupled, so a rise in one leads to a rise in the other, which leads a rise in the first, etc. However, what we are currently seeing is a very rapid rise in Carbon Dioxide levels, and the energy balance model clearly shows that this is a direct cause of temperature rising.

5. Whilst it is of course difficult to predict exactly when the next volcano will erupt, it is possible to make a statistical estimate of the level of volcanic both in the past and in the future. This then can be, and is, incorporated into the climate models.

6. Climate models DO make predictions which can be tested, and they are validated in part by a process of such careful testing. The hind-casting test that we looked at earlier is just such an example. In fact the IPCC reports in [20] and [23] are full of examples of the predictions of climate models compared with actual measured data. Even quite simple climate models have made accurate predictions of the current rise in temperature and sea level.

7. Sadly the *majority* of the comments fall into this category. These are best ignored. You have to develop a thick skin when you work in climate science!

6. Are we past the point of no return?

In Chapter 5, we touched on the theory of tipping points where small changes of the parameters in a model lead to irreversible change in the system. One of the key questions about climate change is whether it is too late to change things, or whether we might be past a point of no return, so that no matter what we do the climate will change regardless.

An example of where this might happen we go back to look at the Energy Balance Model described above and (as we usually do in **Step 7** of the modelling process) we add in additional physics to the earlier climate model. In particular the energy balance model can be improved to account for what is called the *ice-albedo effect*. The albedo a, that we considered as a constant above, in fact depends indirectly upon the average temperature of the Earth. In particular, it depends on the *total amount of ice covering the Earth*. The more ice there is the more reflective the Earth's surface and therefore the higher the value of a. Similarly, a decreases as the amount of ice decreases. The amount of ice in turn depends upon the mean Earth temperature, which from now on we will just call T. The higher the temperature the less ice there is, and the lower the temperature the more ice. Thus the albedo a decreases as the temperature T increases. As a result the sea gets darker and does not reflect the light from the Sun so well when T increases. As a consequence of this, more of the Sun's energy reaches the Earth's surface, and the Earth grows warmer still. In principle this positive feedback cycle could continue until the ice has melted. This is a process of rapid change that would be hard to stop once it has started. This phenomenon is precisely what would be expected at a tipping point.

Is there mathematical evidence for this? It is actually quite hard to model the process of linking temperature to albedo, as we have to take into account the rise and fall of sea ice and also the advance and retreat of glaciers over the Greenland and Antarctic ice sheets, and these introduce delay and uncertainty into the system. If the Earth was very cold and covered with ice it is very 'shiny'; as a result it reflects a lot of the Sun's energy and its albedo would be $a = 0.7$. In contrast if the Earth was very hot and there was no ice at all, then it would be much 'darker' and the albedo is lower. In this case it a mixture of the albedo of the ocean and the land, and would be about $a = 0.29$. A plausible and simple model which takes this into account is to make the albedo $a(T)$ a function of the temperature T which varies smoothly between these two extreme values. An example of a function $a(T)$ which has this property shown in Fig. 6.18 and is given by the formula

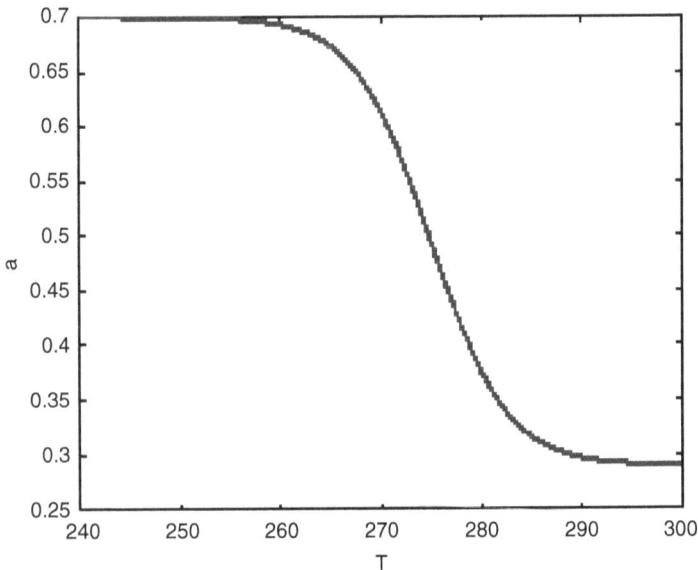

Figure 6.18   *The albedo a(T) varying between a = 0.7 for very cold temperatures to a = 0.29 for very hot temperatures.*

$$a(T) = 0.495 - 0.205\tanh(0.133\,(T - 275)),$$

(where T is measured in degrees Kelvin). This function has the form given in Fig. 6.19.

The effect of making the albedo of the Earth depend upon temperature is twofold.

As a **first** observation this dependence of the albedo upon temperature makes the whole Earth/atmosphere system *much more sensitive to change*. In particular *it increases the sensitivity of T to changes in the emissivity e* which is given by dT/de. As this is in turn linked to Carbon Dioxide levels it means that the change in the temperature due to changes in Carbon Dioxide is amplified. Without the ice-albedo feedback a simple calculation gives

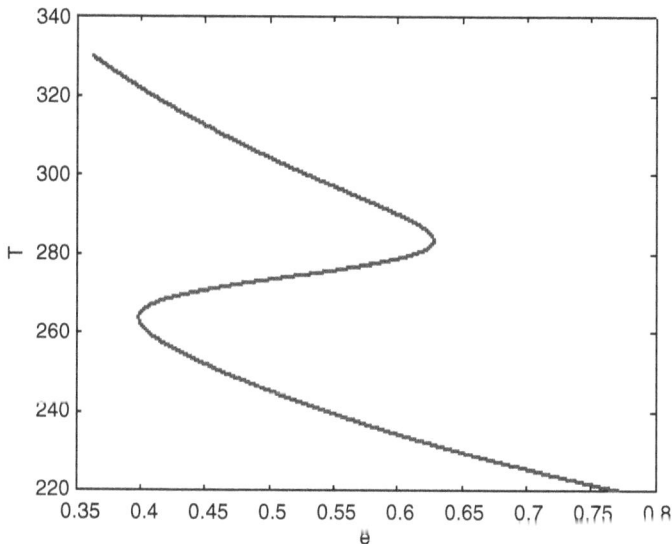

Figure 6.19 *A plot of T as a function of e. This shows the multiple states that the Earth's climate can exist in if the albedo $a(T)$ depends on the temperature. The current state of the Earth's climate has $T = 288$ K when $e = 0.605$. There are tipping points at $e = 0.38$ and at $e = 0.63$ when the climate can change very rapidly.*

$$\frac{dT}{de} = -\frac{T}{4e}.$$

As this is negative, this means (as we have already shown) that as e decreases, then T increases.

If the ice-albedo feedback is included, this sensitivity changes to

$$\frac{dT}{de} = -\frac{T}{4e + \frac{Q\, da/dT}{\sigma T^3}}.$$

Now, as T *increases*, so the albedo a *decreases*. Hence, $da/dT < 0$. This means that the absolute value of dT/de in this case is *larger* than the value calculated above when we did not consider the albedo changes. As a consequence the temperature T increases rather more rapidly for a given amount of Carbon Dioxide added to the atmosphere than it would if there was no change to the albedo of the Earth. This means that our earlier optimistic estimate for the effect of changing the Carbon Dioxide on temperature is simply mistaken due to this feedback effect. It is worth saying that there are many other feedback effects in the climate which can also affect the value of dT/de in both a positive and a negative fashion.

A **second** interesting effect is that in this model the Earth can now exist in a number of different climatic states. For example, if we take the current value of $e = 0.605$, then the energy balance equation has *three different (steady state) solutions* given by $T = 288$ K (our current temperature), $T = 279.6$ K and $T = 232$ K. Of these the first is stable (a warm Earth) as is the third (a cold Earth) and the other value represents an unstable middle state. The multiple states as a function of e are illustrated below in an S-shaped figure. In this figure the top branch is the temperature T of stable warm Earth (the current state) and the bottom branch a stable cold Earth.

This figure shows the existence of two *tipping points* when the climate can change rapidly at $e = 0.38$ and at $e = 0.63$. If we were in a cold Earth situation (often called a *snowball Earth*) and e dropped below 0.38 then we would see a rapid warming of the Earth, similarly if we were in a warm Earth scenario then increasing e above 0.63 would lead to very rapid cooling. It would seem from this model that the current state of the climate is not close to such a tipping point, so we are not past the point of no return (at least in this model). However, the ice-albedo effect certainly makes the whole system much more sensitive to the effects of increased Carbon Dioxide emissions.

There are other aspects of the Earth's climate, however, which might lead to a tipping point in its behaviour. These are well described in the book by Tim Lenton [28]. One of the most commonly quoted of these is the melting of the Siberian permafrost which will lead to the release of the Greenhouse Gas Methane, which will make the Earth warmer, leading to more melting. Other potential tipping points are, the change on the Atlantic circulation, the loss of the rainforests, and the greening of the Sahara.

It is worth saying however that the theory behind the role of tipping points in the climate is still unclear, as is their detection. Nevertheless, they are certainly worth monitoring and this is an area of active research.

7. Some predictions of the future

So, are we all doomed? It is not my job as a mathematician to say this one way or the other. However, I can urge everyone to be mindful of the effects of climate change. Despite what certain politicians may say, the evidence for human-made climate change is very strong. This is supported by mathematical models which imply that unless we do act now the Earth's temperature will continue to rise, and we can make clear predictions about how much that rise will be given the amount of Carbon Dioxide that we are releasing into the atmosphere. How we mitigate that rise, for example by using Carbon Capture Technology, or the use of renewable energy, is the subject of another lecture. But perhaps the main contribution that maths can make is, by using data and careful models, to take the hot air out of the climate debate. If you want to read more on this, I strongly recommend the (free) book written by the late, great, David Mackay [29].

7
Energetic mathematics

1. Introduction

When asked what is the most important invention ever made by human kind, apart from my own personal favourite of calculus (which strangely never gets many votes), one invention that always figures very highly is fire. It was fire that first allowed us to release energy in meaningful amounts. This energy could then be used for cooking, heating, lighting, and the manufacture of new materials such as metals. Since then our whole civilisation has both relied on, and has been defined by, the need to obtain energy. Some of this has come from natural sources such as the wind, the Sun directly, or from flowing water. Until recently the bulk of the rest of our energy came from burning wood (or peat) and then fossil fuels such as coal, oil or gas. The industrial revolution was triggered by the discovery that through the use of the transition of water into steam, the *heat energy* released by this process could be turned into *mechanical energy* and this could be used to power the great machines of the industrial age including lathes, mills, pumps, lifts and, of course, locomotives.

During the 19th Century, the work of Michael Faraday at the Royal Institution in London (where I am the Professor of Mathematics) led to the discovery that this mechanical energy, when coupled to a generator, could lead to electrical energy (and conversely that you could use electricity to power a motor). Through the work of many others, most notably Edison and Tesla, we then saw a revolution at the end of the 19th Century, and the start of the 20th, with the widespread adoption of electricity as the primary means of both transmitting and using energy. Now electricity is generated through a wide variety of mechanisms including wind, solar, hydropower, fossil fuels, and now also including nuclear power, tidal power, wave power, hot rocks and bio-fuels. Of these, fossil fuels currently account for about 80% of the world's energy production, hydropower, wood and nuclear for just under 20% and the remaining 2.5% by renewable sources (although this figure is now rapidly rising). Most of the world uses electricity (although about one billion people have no access to electrical power). The huge advantages electrical energy has over mechanical (and most other forms of) energy is that it can be transmitted over huge distances with almost no loss, and it can be (relatively) easily controlled. This has led it to be widely adopted as the primary source of the world's energy. With the rapid rise in the use of zero-emission electrical vehicles, we are about to see a significant rise in the use of electricity in the future.

The annual consumption of electricity in the UK is about 360 TWh[1] (predicted to rise to 730 TWh in 2050), and the peak demand is around 70 GW, depending upon the time of day and the day of the week. This electrical power is supplied over a complex network starting, usually, with power being generated at a power station. This is then transmitted over a high voltage network, before being reduced in voltage and distributed to commercial, industrial and residential consumers. Maths is vital in ensuring that the lights always stay on as the planners of the grid need to solve a large number of nonlinear differential-algebraic equations, described on a complex network (with 30 million nodes representing different households, industries and other users of electricity), to work out how much electricity can be generated, distributed and stored. However, this is not easy, as electricity must be consumed as soon as it is purchased, it cannot be stored in large quantities and the user has a very low tolerance to interruptions in the supply. These challenges are going to increase significantly in the future with a greater emphasis on low Carbon generation, a much more decentralised supply network (with a significant increase in lower power generation from renewable sources such as solar and wind often at a domestic level), the increase in the use of electric vehicles, an increase in local electricity storage, and the advent of the SMART Grid [34] in which users both have much greater control over their energy demands and also supply much more information to the Grid. For the future planning of the National Grid this raises important questions. For example, how should we expand our power system and what will happen 5, 10 or even 20 years from now (remember that it takes a long time to build a power station, and it will be in use for a long time). Furthermore, where should the generating plants be constructed,

---

[1] A kilo Watt Hour (kWh) is the amount of energy required to supply a kilo Watt (the rough power consumption of a typical household) for one hour. 1 kWh = 3.6 M Joules. Your electricity bill will usually be given in kilo Watt hours, typically around 9000 kWh per year. A TWh is one Tera Watt hour, which is one billion kilo Watt hours.

for both economic and also environmental reasons. In addition, what will be the configuration of the transmission lines, what voltage will they be using and will we see a change from alternating current to high voltage direct current in the future?

In this chapter we will see how the grid system works and the mathematical models which make it work well. I will explain (by using the tipping points that we looked at in Chapter 5) how power cuts can happen (with the example of the US NE coast blackout) and will show how using a mathematical model can prevent this happening in the future. We will then look at the challenges and opportunities presented by renewable energy, the SMART grid and electrical vehicles. There are many challenges for the future supply of energy in as clean a form as possible, and the use of this to power (for example) electrical vehicles. Mathematical models give us a vital tool for addressing them.

2. Where does electricity come from, and where does it go to?

Electricity is mainly generated in large power stations at a very high voltage. In the UK there are around 180 of these power stations. A typical large power station is either burning gas, is coal fired, or is nuclear, and can generate around 2 GW of electrical power. At present the largest supplier of electricity in England at 34.5% comes from gas power stations, and in Scotland by hydroelectric power. Coal fired stations in the UK, once dominant, now only contribute a small amount (if any) of the total electrical power. Nuclear contributes about 17%. The value of 2 GW is similar to the power production of a typical large hydroelectrical plant, but the Three Gorges hydroelectric plant in China produces a staggering 22 GW. A large wind farm generates about 300 MW of power, and wind energy is rapidly increasing in its overall contribution to the total power generated, with wind power amounting to about a quarter of the total energy produced in the UK. In contrast, a domestic solar cell system would produce around 1 kW, although the abundance of solar power now means that it contributes around 5% of the UK's power requirements.

Figure 7.1 *The transmission of very high voltage three phase AC electricity using cables strung from pylons. (Credit: https://www.shutterstock.com/image-photo/silhouette-high-voltage-electric-tower-on-697305988)*

Once produced the electricity is transmitted (in the National Grid in the UK) as three-phase AC (using the method invented by Tesla) at a very high voltage, with the highest voltage being Extra High Voltage (EHV) of 400 kV, over a national network of power cables, typically hung from pylons, as shown in Fig. 7.1. It is then sold onto the regional companies where its voltage is successively reduced, by transformers in sub-stations, to the High Voltage (HV) of 11 kV and then down to the domestic supply Low Voltage (LV) of 415 V and the consumer voltage (in the UK) of 240 V. At all times and places it is at a constant frequency of 50 Hz. In Fig. 7.2, we show the various distributors and consumers.

Producing electricity securely, safely, reliably and cheaply, has many challenges. Electricity is difficult to store in large quantities, so it usually has to be used as soon as it is generated. We also have a very low tolerance to any interruption in the electricity supply. Just think of what happens when we have a power cut. No lighting, no heating, the freezer goes off, and the home Internet shuts

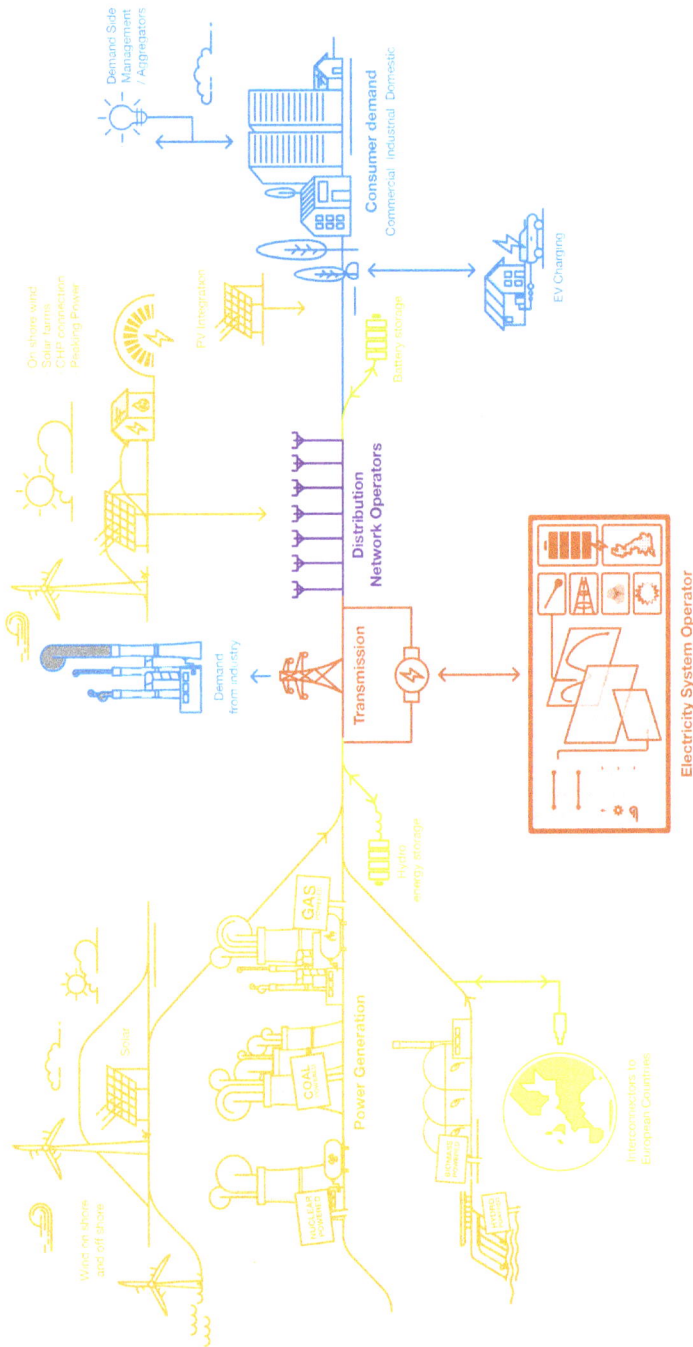

Figure 7.2 *A schematic representation of the electricity supply network showing the different generators and the consumers. (Credit: With thanks to the National Grid ESO)*

down! You want the lights to go on as soon as possible, if not sooner. Other challenges arise from the extreme interconnectedness of the electricity network, which means that a problem in part of the network quickly becomes a problem for the whole network. Most of the time the process of transmitting electricity proceeds smoothly. However, there are times when the users place a very large demand upon the network. An example of this is an international football match when lots of people will not only turn on their TV sets to watch the match, but will also turn on their kettles at half time, or just after the match finishes. In Fig. 7.3, we can see the demands on the UK supply by some notable football matches in recent years. Top of the list was the famous World Cup semi-final of England against West Germany, which went to penalties, but was ultimately lost by England. This match is famous for two things. Firstly, Paul Gascoigne's tears when he received a booking and realised that he would not be able to play in the final. And secondly for nearly shutting down the UK electricity supply network, as described in [36]. Indeed the total change in the demand for electricity during the match was 2.8 GW,

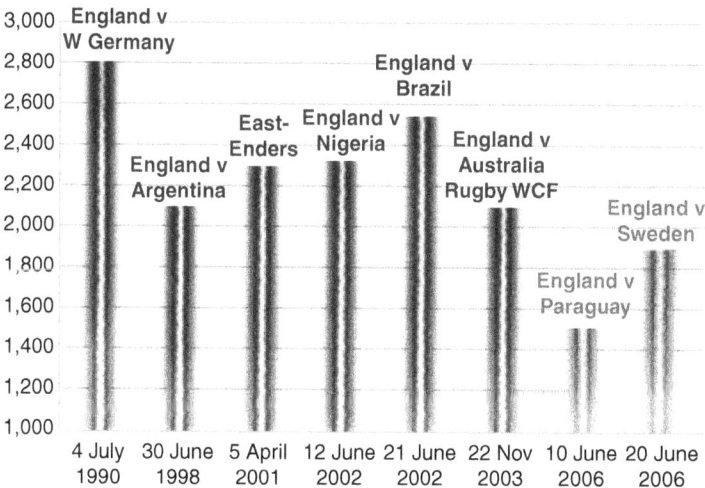

Figure 7.3 *Changes to the demand for electricity in the UK as a result of various sporting fixtures.*

which was 11% of the total power delivered by the network and amounted to about 2 million kettles.

It is the mark of a good electricity supply network (and the operators of that network) that this does not happen, and that the lights always stay on, regardless of the demands placed on the network! Technically this means delivering a secure supply of electricity at a near constant voltage and frequency, at all points of the country, regardless of the amount of power demanded from it. The UK National Grid is very carefully controlled to make sure that the electricity supply remains stable, and the lights have (so far) always stayed on. In the next section we will see why, and the role that maths plays in keeping the lights on.

3. The complex story of how electricity is transmitted and distributed

The war between AC and DC

The modern electricity supply network relies on the invention of alternating current (AC) by Nikolas Tesla. In AC the current and the voltage vary sinusoidally with time as seen in Fig. 7.4, with the voltage varying from a maximum value, to a minimum value and back in one *cycle*.

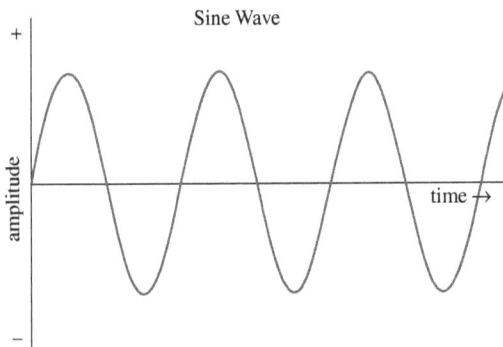

Figure 7.4 *The sinusoidal variation in the voltage of the electricity supply.*

Thus the Voltage $V(t)$ and the current $I(t)$ of the electricity supply have the form

$$V(t) = |V|\cos(\omega t + \varphi_1), \qquad I(t) = |I|\cos(\omega t + \varphi_2),$$

where ω is the frequency of 50 Hz (so that the period of each cycle is 0.02 s) and j_1 and j_2 are the *phases* of each (which we will return to later), and $|V|$ and $|I|$ are the *amplitudes* of each, and t is time.

We have seen this curve before, in a totally different context, when we looked at the low amplitude swings of the pendulum. It is both remarkable, and a tribute to the universal nature of mathematics, that we see the sine wave over and over again in so many diverse applications.

The reason that AC was originally adopted was that it is relatively easy to transform from a high AC voltage to a low one, and vice-versa, by using a transformer. This can be done with a small loss of energy. A high voltage can then be transmitted at a low current, meaning that the power loss on transmission (which is proportional to the square of the current) is then low. Thus AC could be transmitted over large distances at high voltage with little loss of energy.

The reason for this can be shown by using a mathematical model of the power loss and the power generation. This model makes use of two simple expressions. If I is the current through a wire, V is the voltage across the wire, and R is the resistance of the wire then Ohm's law states that:

$$V = IR$$

and the power P of the electricity is given by

$$P = IV.$$

If a wire has resistance R and a current I flows through it then then power loss P_{loss} in the wire (which is useless and only goes to heat up the wire) is given by combining the above two expressions to give

$$P_{loss} = I^2 R.$$

So if you *double* the current through the wire then the power loss *quadruples*.

Now, if the power station produces electricity at a *constant* power P and at a voltage V the current I is given by the formula:

$$I = P/V.$$

So, the higher the voltage at the power station, the lower the current. If we combine these two formulae then we get

$$P_{loss} = \frac{P^2 R}{V^2}.$$

As we have seen, a typical domestic voltage in the UK is 240 V and the typical voltage of a high voltage power line is more like 400 kV. Using the above formula the ratio of the power loss of the high voltage line to that of a domestic line is $(240/400{,}000)^2 = 1/2\,777{,}777$. So using the high voltage reduces the power loss by a factor of nearly 3 million. The mathematical model therefore shows that it makes huge sense to transmit electricity at a high voltage. Only when it is needed to supply a domestic household is it necessary to transform the electricity down to a lower voltage. This means that electricity can be generated at high voltage efficiently by a large power station far away from a city and transported with very little power loss into the city to be used. In contrast low voltage generation needed lots of inefficient local power stations. In the early part of the 20th Century a bitter battle was waged between Tesla (working for Westinghouse) and Edison, about the best way to produce and transmit electricity. Edison favoured low voltage direct current (DC) in which the voltage is constant, claiming that it was safer. (Edison tried to demonstrate this by advocating the use of electricity in the electric chair to execute convicted criminals.) Edison built some small local power stations to supply electricity. However, as we could have predicted from the above model, there was a huge energy loss between these power stations and the users. The advantages of AC so greatly outweighed those of DC, that it became widely adopted and is still very much in use today. It is worth saying that recent advances in power supply design mean that it is possible, by using controlled switching devices called

buck convertors, to transform a high DC voltage to a low one with minimal energy loss. This has led to a recent comeback of high voltage DC or HVDC, and this is now being used for the power cable link between England and France.

Complex numbers and their applications

To represent an AC voltage, electrical engineers make extensive use of *complex numbers*. The use of complex numbers is an amazing example of where a piece of mathematics, created to solve the 'pure mathematical' problem of finding the roots of a quadratic or a cubic equation, has found boundless applications.

If you multiply a positive number by itself, then you get a positive number. Similarly, if you multiply a negative number by itself then you also get a positive number. So at first sight it doesn't seem to make any sense to ask if the quadratic equation

$$z^2 = -4$$

has a solution. However, this little difficulty doesn't stop a creative mathematician. They can speculate about what sort of number might satisfy this equation, and then try to discover the properties of this number. Later they may even find a use for it! A huge amount of mathematics works this way, which is one reason why we should always support the creation, and research in to, seemingly abstract mathematics.

Exactly these considerations led the mathematicians of the 18th Century to introduce the idea of *imaginary numbers* which allow a solution to this equation. The imaginary number i satisfies the equation

$$i^2 = -1.$$

Using these numbers we can then solve the original equation $z^2 = -4$ to give the answer

$$z = 2\,i.$$

A number of the form $y\,i$ where y is a 'normal' or a 'real' number is called an *imaginary number*.

A *complex number* z is a combination of a real number x and an imaginary number $y\,i$ to give

$$z = x + i\,y.$$

In this expression for the complex number z we call x the *real part* and $i\,y$ the *imaginary part*.

You can add, subtract, multiply and divide complex numbers with the following rules:

$$(a + b\,i) + (c + d\,i) = a + c + (b + d)\,i,$$
$$(a + b\,i) - (c + d\,i) = a - c + (b - d)\,i,$$
$$(a + b\,i)(c + d\,i) = (ac - bd) + (ad + bc)\,i,$$
$$(a + b\,i)/(c + d\,i) = ((ac + bd) + (bc - ad)\,i)/(c^2 + d^2).$$

For example, $(2 + 3i) + (4 + 6i) = 6 + 9i$, and $(2 + 3i)(4 + 6i) = -10 + 24i$.

You can do many, many things with complex numbers. You can use them to solve polynomial equations, they play a very important role in geometry (and hence computer graphics) and you can do calculus with them such as integration and differentiation. Indeed, a lot of seemingly very hard integrals become very easy when you use complex variables to solve them.

Complex numbers were originally thought to be highly abstract mathematical objects of no possible use. They are now known to be hugely important in a vast number of applications ranging from Quantum Mechanics to the study of water waves, and from vibrations in engineering structures to animating movies. In particular, they lie at the heart of the mathematical models of power engineering.[2]

---

[2] In power engineering the convention is usually to use j to represent an imaginary number to avoid confusion with current denoted by i. However, in mathematics it is conventional to reverse

The reason for this is famous identity which was discovered by the amazing Swiss mathematician Leonhard Euler in the mid 18th Century, and which states that:

$$e^{i\theta} = \cos(\theta) + i\,\sin(\theta).$$

This identity is without question the most important formula in the whole of mathematics!

Whole books have been written about it, see for example [38] and [39].

Its sheer elegance is often illustrated by taking the special case of taking $\theta = \pi$. In this case $\cos(\theta) = -1$ and $\sin(\theta) = 0$, which leads to the identities:

$$e^{i\pi} = -1 \quad \text{and} \quad e^{i\pi} + 1 = 0.$$

These identities unite all of the main numbers in mathematics into one formula. If there was a contest for the best ever formulae then this would win every time! We will see it in application many times for the remainder of this book. When I was a teenager I had a T-shirt with the first of these formulae on, which I proudly wore around town. Happy days!

Euler's theorem has countless applications, and in this chapter we will see how it is used in power engineering.

The reason in this case is that it allows us an easy way of describing alternating current, along with its frequency and phase. Using this an alternating voltage, as described above, can as the real part of the function

$$V(t) = |V| e^{i(\omega t + \varphi_1)}$$

---

the situation and to use i for the imaginary number and j for current. It is pointless to argue which is better (or worse), but it is a shame that this separation exists between mathematics and engineering. Without making any judgment one way or the other as to which is best, I will use the mathematical convention in this chapter. However, I'm instantly happy to switch my allegiance when I'm working with power engineers. Indeed, not so long ago, I was one myself.

with a similar expression for the current. Here, $|V|$ is a constant real number called the modulus or the *amplitude* of V. If $z = x + i\,y$ then

$$|z| = \sqrt{x^2 + y^2}.$$

A convenient way to express this is as

$$V(t) = |V| e^{i\varphi_1}\, e^{i\omega t}$$

and we call the expression $V = |V| e^{i\varphi_1}$ the *complex voltage*.

This single complex number contains two pieces of information, namely the amplitude $|V|$, and the phase φ_1, of the voltage. There is a similar expression for the *complex current*. Using this we can now describe how a power network functions and how we can use maths to make sure that it operates well.

4. What a power distribution network looks like, and a mathematical model for its operation

In any power distribution network (a small one is shown in Fig. 7.5) we have supplies of electrical power from a power station, many households which are supplied by power from the network, and junctions called *buses*. Each bus will be at a particular voltage. Between the buses there are connections. These can be the high voltage cables that we see proudly marching across the countryside. The cables carry current between the buses, which, as we have seen, is kept low by having a high voltage. Low current leads to much smaller power losses, which is why high voltages are used.

In a typical network there are many buses, which can be power stations, factories, transformers, switches, points where the network changes, and households. In principle there can be one bus for every household, so up to 30 million buses. This immediately gives you some idea of the scale and the complexity of the electricity supply network. At each bus the network needs to supply a certain amount of power. How much depends upon the usage and load imposed on the network. We calculate this as follows. Each bus is numbered by an index

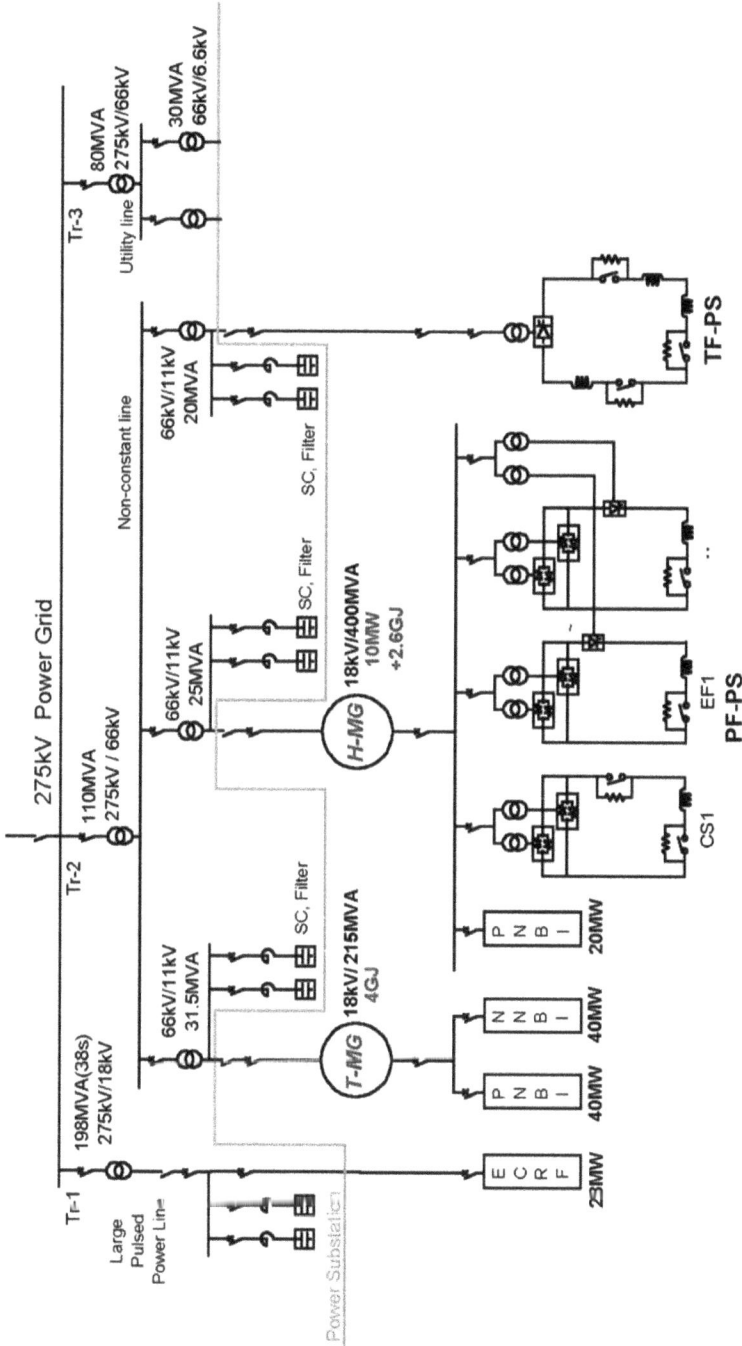

Figure 7.5 An illustration of a part of the electricity supply network, showing the suppliers and users of electrical power.

$j = 1, 2, 3, \ldots N$ and will have a (complex) voltage V_j. This bus will in turn be connected to many other buses in the network. Typically a (complex) current $I_{j,k}$ will flow between the buses labelled j and k in the network. The *power S* of this current flow is given by

$$S_{j,k} = V_j I^*_{j,k}.$$

Here $I^*_{j,k}$ is the complex conjugate of the complex current. If $z = x + y\,i$ is a general complex number then its complex conjugate is given by

$$z^* = x - y\,i.$$

The total power at the jth bus is then given by

$$S_j = \sum_{k=1}^{N} S_{j,k}.$$

This power is in turn a complex number and we write this as

$$S_j = P_j + iQ_j.$$

Here P_j is called the *real power*, which is the power transferred from the power station and which does work, such as heating your home or running your washing machine.

In contrast Q_j is called the *reactive power* and this doesn't do any work. It is the portion of the total power which returns to the power station in each cycle.

Electrical engineers take the reactive power into account when designing and operating power systems, because although the current associated with reactive power does no work when it gets to you, the user, it must *still be supplied by the power station*. Failure to produce sufficient reactive power to the electrical grid can lead to lowered voltage levels and under certain operating conditions (as we shall see in the next section) to a complete power blackout.

If the voltage difference between two buses is $V_k - V_j$ then the current $I_{j,k}$ flowing between them is given by Ohm's law. In particular there is (another) complex quantity called the conductance $\sigma_{j,k}$ so that:

$$I_{j,k} = \sigma_{j,k}(V_k - V_j).$$

In high voltage cables the conductance is usually close to being a purely imaginary number. This is because these cables are designed to have a very low resistance, but usually have some inductance.

We can now combine the three equations above for the power supply network and we find that the total power supplied to the jth bus is given by the expression:

$$S_j = \sum_{k=1}^{N} V_j \sigma_{j,k}^{*} \, (V_k^{*} - V_j^{*}).$$

Now, when designing and controlling a power supply network, we know the values of the real power P_j and of the reactive power Q_j that needs to be supplied to each bus. This is because the buses represent different types of users, such as households or factories, and we know the usual power demands of each of these. The voltages at each bus then satisfy the equation:

$$\sum_{k=1}^{N} V_j \sigma_{j,k}^{*} (V_k^{*} - V_j^{*}) = P_j + i Q_j.$$

We have finally arrived at our desired *mathematical model for our power supply network*. It relates the voltages at each bus, to the real and reactive power needed by each bus, and the conductance of the power cables joining one bus to another.

If we can solve this equation then we can work out the voltages at each bus which are needed to supply the desired amount of real and reactive power.

This allows a power engineer to be able to simulate the network. In particular to make sure that it can supply enough voltage (and current) to meet the demands of the users of the network. It also tells the engineers which power stations need to ne 'on-line' at any one time, and solving it is therefore vital if we are going to get the electricity that we need to keep our houses and factories functioning.

Take a look at the equation. It is nothing other than a large collection of **quadratic equations** for the (complex) voltages (meaning that it is made up of products of one voltage with another). We can therefore understand the complex behaviour of the grid by looking at the complex solutions of quadratic equations and we will do this in the next section.

It is worth noting that an attempt was made a few years ago in the UK to stop the teaching of quadratic equations at school on the grounds that they were totally useless, of no possible benefit to the education of anyone, and that teaching them would simply frighten the students. We illustrate this terrible situation in Fig. 7.6. This suggestion led to a debate in the House of Commons in the UK in which the merits and disadvantages of teaching the quadratic equation were discussed by MPs. Possibly the first time that an equation has been the subject of a debate! More details of this debate are given in the account by Budd and Sangwin [31]. Fortunately the outcome of this debate was that schools should continue to teach their students about quadratic equations and how to solve them. This is fortunate as we need to solve quadratic equations if we want to keep the lights on!

Figure 7.6 *The scary quadratic equation which is actually very useful. (Credit: Plus Magazine [31])*

5. Power cuts and tipping points

Usually, thanks to careful planning by the National Grid, and other energy companies, the lights do usually stay on. This is due to them being able to solve rapidly the mathematical model equation above and to follow the solution as the demand on the power grid varies throughout the day (including during football matches).

Sadly, however, this doesn't always happen. On 14 August 2003 a catastrophe hit the North East coast of America [35]. During a storm, an overgrown tree hit a power cable. The safety mechanisms then cut in leading to a local shutdown of the power supply to the part of the network closest to the affected cable. Unfortunately, there was a software error in the control room, which meant that the shutdown spread and spread. As a result the whole of the North East of America and large parts of Canada were plunged into darkness. The resulting blackout lasted for about two days, and affected an estimated ten million people in Ontario and 45 million people in eight US states. The event contributed to at least 11 deaths and cost an estimated $6 billion.

So, could this happen again? One of the main reasons for a power failure is a phenomenon called a *Voltage Drop* when the voltage in the power network rapidly drops to zero. The cause of a Voltage Drop is essentially one of the *tipping points* which we looked at in Chapter 5. As we might expect from our model, this follows from properties of the solutions of a quadratic equation.

Usually a quadratic equation is written as

$$ax^2 + bx + c = 0$$

and we want to find the solution x. When we meet the equation at school we are told that it may have one, two or no real solutions. These are given by the famous *quadratic formula*:

$$x = \frac{-b \pm \sqrt{b^2 - 4ac}}{2a}.$$

The quadratic formula has an interesting history which I explain further in my article [31]. Quadratic equations have been studied since the times of the Babylonians and are recorded on Babylonian cuneiform tablet. (This is because they are related to *areas* and if the areas are those of fields then they become related to land values, and hence to *taxes*.) The original solutions of the quadratic equation were obtained geometrically. However, the Indian mathematician Brahmagupta (597–668 AD) derived the algebraic formula above for the solution, which he described using words. The first publication of the quadratic formula was in 1637 by the French mathematician/philosopher René Descartes (1596–1650) in his book *La Geometrie*.

The quadratic formula says that the quadratic equation has *two real solutions* if $b^2 > 4ac$ and *no real solutions* if $b^2 > 4ac$. Instead, if $b^2 < 4ac$ then it has two *complex solutions*.

To illustrate this we will look at an equation for which $a = 1$, $b = -2$. The two solutions are then given by:

$$x = 1 \pm \sqrt{1 - c}.$$

There are two real solutions if $c < 1$ and no real solutions if $c > 1$. A graph of the two real solutions is given in Fig. 7.7, with the special point $c = 1$, $x = 1$ highlighted.

We have seen this point before in Chapter 5. It marks the boundary between when the quadratic equation has two solutions and none and it is a *tipping point* for the dynamical system

$$\frac{dx}{dt} = ax^2 + bx + c.$$

In particular if $c < 1$ then this equation has two fixed points, one of which is stable and the other is unstable. If $c > 1$ then the dynamical system is completely unstable.

The link between this and the model equation for the power network is as follows. The system of quadratic equations in the model replaces the single quadratic

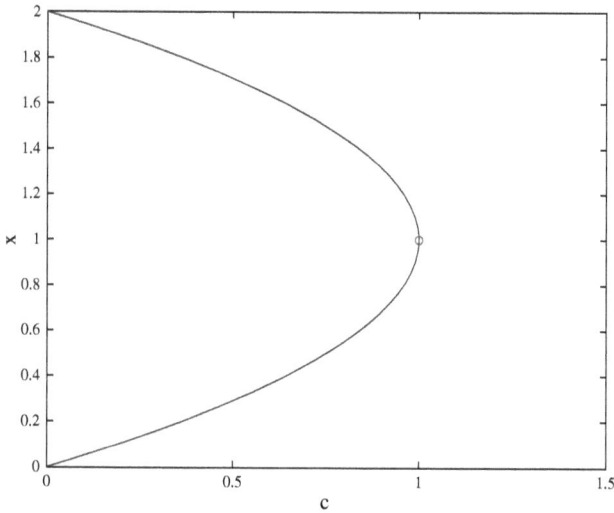

Figure 7.7   *The solution x of the quadratic equation as a function of the parameter c. If c < 1 then there are two real solutions, if c = 1 there is on real solution, and if c > 1 the solutions are complex. There is a tipping point at c = 1.*

equation above. The differential term *dx/dt* above is replaced by a similar term which represents the dynamics of the power grid (due for example to the speeding up and slowing down of the generators in the power stations, and the motors in industrial plants). The term c in the quadratic equation becomes the power drain in the network. If the power drain is low then the whole system can operate in a stable state which changes as the power drain changes. However, if the power drain is larger than the network can manage to supply, then it becomes unstable, and there is a Voltage Drop at the tipping point.

Figure 7.8 shows the normalised (with respect to a usual operating voltage) average voltage in a network when it has to supply a reactive power *Q* to a city. This curve is obtained by directly solving the model equation for the power grid. The three curves correspond on the left to a network with fewer power stations on and thus less available power, and on the right to one with more power stations on and thus more available power. Each of these curves has a *very similar shape indeed* to that of the solution of the quadratic equation illustrated above, complete with a tipping point. This is of course no coincidence

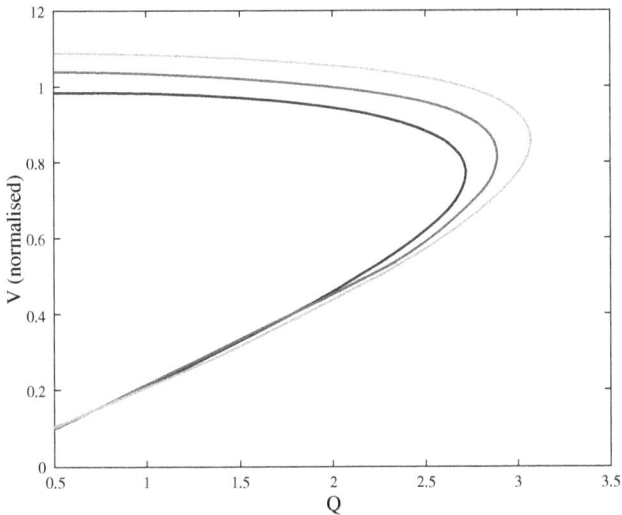

Figure 7.8 *The 'nose curves' of the electricity supply, showing the normalised voltage V as a function of the reactive power Q. The curve on the left corresponds to having fewer power stations than normal, and the one on the right to having more than normal. If Q exceeds the value at the 'nose' then the voltage collapses, leading to a power cut.*

as each such curve is given by solving several quadratic equations instead of just one. These curves are given the 'technical name' of *nose curves* for reasons which should be obvious.

Each nose curve has, as we might expect, two solutions for a given value of Q before the tipping point. The *upper part* of each of the curves represents (as in the case of the quadratic equation based dynamical system) a stable solution in the normal voltage range, which is where we want to operate the power grid. As the reactive power drain Q increases, so the average voltage V drops, slightly at first, and then more rapidly as the tipping point is reached. When we get to the tipping point there is no solution at all. What happens at this point is that the grid goes into a very unstable state during which the *voltage collapses* and all of the lights go out. It is fair to say that, due to good planning and management by the National Grid Company, the UK has never experienced a widespread voltage collapse. However, it has occurred in both Italy and Sweden as well as in the northeast of the USA, and all because a quadratic equation didn't have a real solution.

There are other causes of power cuts, for example on 13 March 1989 the particles emitted from a large solar flare so overwhelmed the safety systems of the Quebec power system that a power cut resulted. On the same day, also due to the solar flare, there was a beautiful display in the UK of the Aurora Borealis which I remember well. (It was my wife's birthday and we thought that this was a lovely birthday present for her.) Quebec also suffered some years later from a major power cut caused by an ice storm which made all of the power cables break with the weight of the ice on them.

6. The future of energy and its control

One of the biggest challenges faced by the energy industry is the transition to a low Carbon technology and a move away from a reliance on fossil fuels (and possibly from nuclear fuels as well). Essentially this means a transition to renewable forms of energy such as wind energy, solar power and other forms of power production such as wave and tidal power. Table 7.1 from a report published by the (then) Department of Energy and Climate Change (DECC) in 2010 [32] shows some possible predictions of future energy production (in

Table 7.1   *The projected power in Giga Watts generated by various forms of energy supply. Note the large amount generated by renewables in 2050. (Reproduced with permission from F. Li [32])*

GW	2009	2030	2050
Wind and wave	1.9	65.9	93.3
Solar	0.0	5.8	70.4
Other renewables	1.8	3.3	3.7
Nuclear	10.9	16.4	40.0
CCS coal	0.0	10.2	39.0
Gas	32.6	28.3	0.0
Coal	23.0	1.3	1.3
Oil	3.8	0.0	0.0
Hydro	1.5	1.1	1.1
Pumped storage	2.7	2.8	2.8
Total	**78**	**135**	**252**

GW). Note the high level of wind and solar and also CCS Coal (Carbon Capture and Storage burning of coal).

Whilst the renewable sources have significant benefits in the reduction of Carbon Dioxide production, they present a big challenge to the grid. In particular they are an intermittent source of energy, and the amount of energy available from them will always be uncertain. In particular both wind and solar energy are dependent in the short term upon the weather, and in the long term by (amongst other factors) the effects of climate change. This predicted change in the way energy is supplied is combined with a projected significant increase in electricity use, particularly for transport, as we move away from internal combustion engines to electrical vehicles. To meet both the supply and demand problems we will increasingly rely on a diversity of different sources of energy production with (possibly) the majority produced by renewables, but with back-up supplies (probably from fossil fuels combined with nuclear energy) and significant redundancy in the network needed to maintain the security of the system. All of this places much greater demands on the control and supply of electricity in an optimum cheap and safe way, and this requires careful mathematics. In part this will be achieved by encouraging the users of electricity to be 'smarter' in their use, with SMART meters giving constant reports on electricity usage by individual households. However, this does in turn lead to additional mathematical problems. At the moment the EHV (Extra High Voltage) supply network is modelled in computers by solving the quadratic equations in the model system above for several thousands of points, in which the electricity demands of a whole town and indeed the HV (High voltage) and LV (low voltage) components of the grid may be modelled as a single point. This is done to allow the calculations to be performed in a reasonable time, so that the behaviour of the network can be predicted over the short time-scales needed to control it, and it is accurate enough in the context of large amounts of electricity being supplied by a small number of big power stations. However, the increased complexity and volatility of the supply (with, for example, many individual households supplying an intermittent amount of electricity to the grid through solar panels on their roof) means that this approach is no longer accurate and electricity supply companies are finding it harder to predict the demands on the grid. There is consequently now a move to develop simulators, which model every single household. This means solving (for the UK) over 30 million coupled quadratic equations every five minutes. Remember, that school children were being warned about the health effects of just solving one such equation! Mathematically solving this very large system quickly is a very challenging task, even for a supercomputer, and developing effective methods to do it is the subject of ongoing research.

One reason solving this large system is so hard is related to the changing nature of the electricity supply. As I described above, we now have a multitude of different suppliers of electricity, all of which has to be added into the National Grid. Some of this, for example solar energy, or energy stored in batteries, is generated as DC rather than AC. This needs to be converted to AC at the right voltage and, crucially, at the right frequency and phase, to patch into the grid. This is achieved using devices called power inverters, which use phase locked loops to synchronise the generated supply to the mains. This introduces additional dynamics into the grid, with the possibility of additional instabilities and lack of security of the supply. To study this we augment the quadratic equations above with additional *differential equations* and study these using the mathematical theory of dynamical systems that we talked about in Chapter 5. As we saw in this chapter, under some circumstances dynamical systems can have chaotic solutions. It is an intriguing, and disturbing, question as to whether the grid could behave chaotically under certain conditions, and some possibilities for this happening are described in [33].

7. The answer IS blowing in the wind

Perhaps the most promising (new) source of renewable energy at the moment is wind power. We can see in Table 7.1 that it is predicted to become *the* dominant source of energy by as early as 2030. A large off-shore wind farm can produce large quantities of electrical power, in the order of 500 MW, and wind power can in principle, be available all day and night, most days of the year. Mathematics is used in a number of ways to make wind power more effective.

Figure 7.9 *The Thanet offshore wind farm. (Credit: https://www.shutterstock.com/image-photo/ view-on-london-array-offshore-wind-1952137951)*

A set of turbines from the Thanet offshore wind farm is illustrated in Fig. 7.9.

These turbines are in a very harsh environment. As well as the forces from the wind, which are required to turn the blades of the turbine, the turbines also have to cope with the force of the sea. In particular, as well as responding to the 'average' sea and wind conditions needed to produce electricity, they also have to deal with extreme conditions, including the intense '100 year storms' which may well occur in the lifetime of the wind farm. This requires careful calculation, both to construct strong enough structures able to deal with the forces involved, and also to predict the nature of the 100-year storms. The latter requires the use of the mathematical/statistical theory of *extreme events* which we also looked at in the last chapter.

Large wind farms are built offshore in part because that is where the strongest and most consistent wind is to be found. Onshore wind farms have to be very carefully sited to receive anything like the same amount of reliable wind.

Unfortunately some of the best locations for these farms, from the point of view of producing energy, are also some of our most beautiful and/or environmentally sensitive areas. Building a large wind farm there would be at best highly controversial and counter productive. It follows that it is essential to maximise the potential of those wind farms which can be built on onshore sites.

There are many ways in which mathematical modelling can help in the design, control, and operation, of a wind farm. The power generated by a wind turbine is mainly determined by the average wind speed. Higher wind speeds rotate the turbine blades faster, meaning that they can deliver more power (unless the wind speed is so high that the turbine has to be shut down).

The model equation for wind power P (in Watts) in a wind of speed V (in metres per second) is given by:

$$P = \frac{1}{2}\rho A C_p E V^3.$$

Here ρ is the density of the air, A is the area of the turbine, Cp is the coefficient of performance (related to the shape of the blades) and E is the efficiency of the turbine gearbox and generator.

The reason for the cube of V in this equation is that the power per unit area of the wind is proportional to its kinetic energy times its momentum. This formula predicts that if the wind speed doubles, then the power extracted from it increases by a factor of eight. This is an enormous increase, and one of the reasons that wind power is such a promising source of future energy. The model equation for wind power allows us predict how much power the wind farm will deliver *if* we can predict the wind speeds.

It is also interesting to ask why most wind turbines have three blades. The reason is twofold. The more blades that a wind turbine has, the heavier it is and the more air resistance there is to move the blades. So it makes sense to have as few blades as possible. However, if a turbine has only two blades then it can easily become unbalanced when there is more wind loading on one blade than the other. The optimal solution is three blades, which is more stable than two because three blades better spreads the wind load, and has less air resistance than four blades.

Wind turbine blades are typically glued together. Another mathematical model, then allows us to calculate what the maximum wind loading is on the blades. This gives information to the manager of the wind farm on what wind speed means that the wind farm needs to be shut down to keep everything safe.

The above discussion means that to know both what electricity supply from the wind farm is likely, and how it is best matched to the demands of the grid, and also when the wind farm may need to be shut down, then we must solve the seemingly impossible task of accurately predicting the wind speed at the precise location of the wind farm. As an example of the magnitude of this challenge, Fig. 7.10 gives a plot of the seemingly very erratic wind speed given hourly at a single location over a period of two years [37].

In Chapter 6, we looked at the mathematics behind weather forecasting in which the wind is predicted about 5 days ahead, on a grid with a resolution of

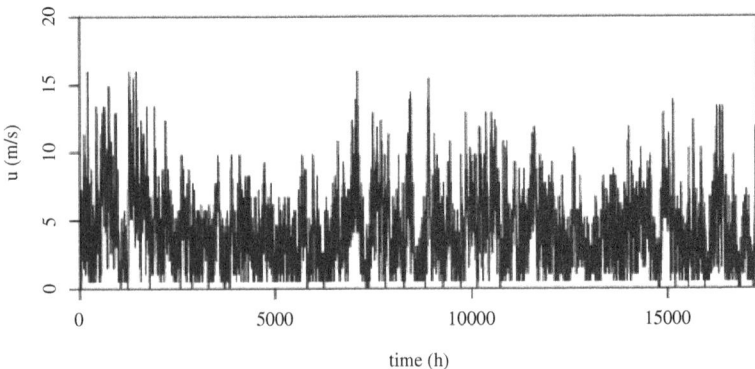

time (h)

Figure 7.10   *The erratic speed of the wind over a series of hours.*

about 1.5 km. Unfortunately for a wind farm this prediction is unlikely to be accurate enough for their specific location.

Nevertheless, mathematical help is at hand. What the wind farm and power company managers often need to know is not what the wind will be doing at their location tomorrow, but instead what it will be doing in the *next few hours*. Furthermore, the managers have a lot of data at their disposal, namely the wind speeds at their precise location, which they need to know to be able to run the farm at all (and which they can measure directly from the wind turbines). This question of the short-term forecast can then be tackled by using statistical and *machine learning methods*. These methods take the great mass of wind data at the location gathered over a period of several months, and then train a statistical (typically an *auto regressive moving average* ARMA model) or a machine learning model (typically a neural net) on the data. These statistical or machine models are highly effective in making short-term predictions of the wind local to the site of the wind farm. In fact over the next four hours they will typically out-perform a more traditional weather forecast. In much the same way that the most accurate way to tell what the weather at your house is *right now* is simply to look out of the window. Because of this, statistical and machine learning models are now being used very effectively by the wind-power industry. It is the continued use of mathematics in this way which is helping to make wind power a viable source of mass energy in the future.

If you want to know the weather *tomorrow* then a traditional forecast based on a mathematical model such as the one described in Chapter 6 is currently more accurate. It remains to be seen whether in the future the use of statistics or machine learning will replace more traditional models for the five-day forecast.

8. Conclusions

Energy matters to all of us, and the challenge of supplying enough energy in a clean and safe way, is one of the greatest challenges faced by humanity. Energy supply networks are already highly sophisticated, and will be more so as the way in which we use and supply electricity is changing rapidly. To deliver a secure supply well into the future will require equally sophisticated mathematics. However, the bottom line is, that to keep the lights on we all need to be able to solve (lots of) quadratic equations. For many of the other uses of a quadratic equation see the *Plus Magazine* article [31].

8
Mathematical food for thought

1. Introduction

A universal constant amongst all animals is the need to eat. The appreciation of food and drink has been one of the greatest forces moulding our lives both from a point of view of day-to day-existence, and also our sense of taste and aesthetics. Food truly brings us all together, and without it we would all surely die. The food and drink industry is the largest in the world and in order to feed the growing population of the world we will have to grow more food in the next 50 years than we have in the last 10,000 years.

In this chapter we will look at the role that mathematical models play in the production, cooking and consumption of food and drink, taking you from 'farm to fork' with a series of case studies based on my own experiences. I hope that you will all enjoy this rich diet of mathematics.

2. Some basic facts

What does mathematics do to help the starving people in Africa? This is a question I am not infrequently asked when I give talks to schools, where the pupils are not aware of the many applications of maths to the real world. The simple answer to this question, is that mathematics in general, and mathematical models in particular, help a very great deal, by ensuring that the starving people will be fed. Let us think about the various processes involved. In the case of arable farming this involves planting the seeds, watering them and letting them be pollinated and then grow. It may be also necessary to apply pesticides at some point, and to understand the weather well enough to know when to harvest. Following harvesting the food must be transported to where it is needed. Other types of food, such as cattle, pigs or chicken, must be raised carefully and allowed to grow. Chicken eggs need to be incubated and foodstuffs for the animals need to be delivered on time. Once the animals have been slaughtered for their meat, it is usually refrigerated and stored. This has to be done very carefully to make sure that it is safe, and that the meat will be fresh when it is thawed out. This meat also needs to be delivered to the customer in such a manner that it does not lose its freshness. Once at the consumer the food must be cooked safely, eaten and finally digested. Drinks must also be prepared carefully. Mathematics plays a vital role in ensuring that we have safe water to start with. It helps brewers in controlling the fermentation and production process, and in storing beer to avoid sedimentation. It also, as we shall see, helps to put the bubbles in beer and stout.

Sometimes when I talk about food, I am told off for using mathematics and science on 'something so trivial'. The government disagrees. As a measure of its importance, agricultural science is listed as one of the UK Government's *Eight Great Technologies* in a list published in 2012. The UK agri-food industry alone contributes around £100 billion annually to the economy. As part of this the drink industry contributes £18 billion, with 5 billion pints of beer drunk per year (which works out as two pints per week per head of the population). Food related companies and large online retailers employ mathematicians, sometimes in large numbers. Mathematicians are also in demand well before the food reaches the shop shelves. As an example, mathematicians work at the heart of the chocolate industry. It is hard (for me) to think of a better occupation than being a chocolate mathematician.

3. It all starts in a field

Apart from fish (which we will come back to later), most of our food production, whether it is crops or animals, involves a field on a farm (or its close relative and orchard or a vineyard). This leads to the often quoted phrase *food from farm to fork*. Whilst it might look on first inspection to be a low technology item, a lot of science and mathematics is involved in making a field effective for food production. Indeed there are sophisticated computer packages which are used to simulate the behaviour of a field and to advise farmers on the best way to manage the fields on their farm. The basic questions that need to be addressed by a farmer growing crops are: what crops to grow and how much, how much to irrigate them, what pesticides should be used, how to react to the weather and when to harvest. The first of these questions involves the mathematics of optimisation. It is perhaps useful to give an idea of how this might work with a simple field (or maybe several fields) on a farm.

Let's consider an actual example of a farm somewhere in the tropics in which we want to grow two crops, such as cocoa and pineapples. We suppose that c is the amount of cocoa we plan to grow in one year, p is the amount of pineapples, and the unit cost of cocoa seed is a, and of pineapple seed is b. A simple mathematical model of the total cost C of growing the two crops is given by:

$$C = a\,c + b\,p + d.$$

Here d is the upfront cost we must take on just to use the field in the first place. This includes costs such as the rental for the field, labour, irrigation, pesticides, insurance, etc.

Similarly, when we harvest the crops we might expect a unit return of e on the cocoa, and of p on the pineapples. Thus we might make a profit P given by:

$$P = e\,c + f\,p.$$

Finally, if the amount of space taken up by a unit cocoa is g and by a unit pineapple is h then the total amount of space S taken up in the field by our two crops is given by:

$$S = g\,c + h\,p.$$

The problem faced by our farmer is then as follows. They want to grow the right amount of cocoa c and pineapples p which in turn *maximises* their profit P. But at the same time they must also want to keep the cost C below some maximum C_{max}, (their total available cash) and require that the space taken up is less than the total area of the field given by S_{max}. These two conditions are called *constraints*. In mathematical terms the problem of maximising the profit becomes:

Maximise: $P = e\,c + f\,p$ over all *positive* values of c and p,

subject to: $0 \le C \le C_{max}$ **and** $0 \le S \le S_{max}$.

The problem of maximising the profit, is a special example of *a constrained optimisation problem* called a *linear programming problem*. In the case of our problem above, this can be solved by using a graph. Each of the constraints defines a region which is bounded by a straight line. The points which satisfy all of the constraints then lie in a polygonal region which is bounded by these lines. In Fig. 8.1, we show an example in which the optimal combination of c and p is highlighted. It occurs at the corner of the shaded four-sided figure bounded by the axes and the lines defined by the two constraints. The shaded region which satisfies all of the constraints is called the *feasible region*. The optimal solution which *maximises the profit P* is given by the 'corner' of the feasible region which in turn is given by the intersection of the lines defined by taking equality in each of the constraints, so that we take $C = C_{max}$ and $S = S_{max}$.

More generally of course, a farmer will have many more crops, or even animals, to consider, as well as many more constraints, such as labour costs, irrigation costs, etc. as well as considering the resistance of each crop to disease. This leads to more complex problems similar in form to the problem above involving many more variables and constraints.

Not dissimilar problems arise in many other applications, such as in retail and in transport systems. A key feature of the problem above is that it is *linear*. This follows directly from our very reasonable modelling assumptions, which means that we see c and p and their multiples in it, but not more complicated functions such as c^2, c^3 or c^p. We call such a problem a linear constrained optimisation problem.

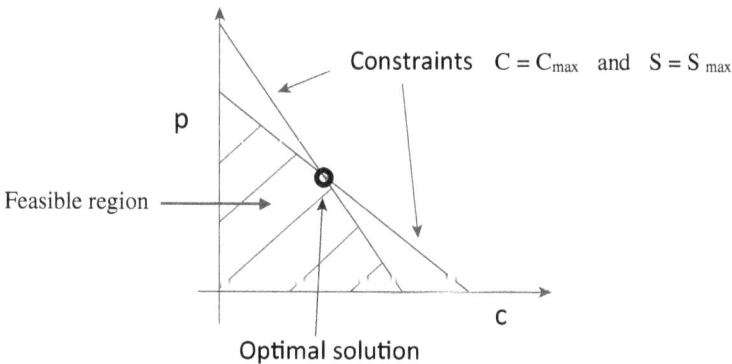

Figure 8.1 *The optimal solution is found by taking the intersection of the two straight lines defining the constraints.*

Remarkably, despite their apparent complexity, there is an algorithm to solve all such linear problems. It is called the Simplex Algorithm, and its invention in 1947 by Dantzig was one of the key developments in mathematical algorithms in the 20th Century. Today countless optimisation problems are solved by the Simplex Algorithm, ranging from farming to some of the most complex problems in economics and scheduling. More information about the Simplex Algorithm can be found in the book [40] and the Wikipedia article [41].

Things get a bit more complicated when we have to consider the effects of weather and climate. In Chapter 6, we looked at the mathematics behind this complicated subject, but it is obvious (and has been for eternity) that the weather has a huge effect on the productivity of a farm. As an example we can look again at the production of cocoa. In Fig. 8.2, we show the total yield of cocoa in Nigeria each year and compare it with the mean rainfall per month. This shows a clear link between the two. More details of the link between cocoa growing and climate are given in [42].

Farmers must thus plant their crops taking into account the predicted effects of the weather. In the past this was largely a matter of luck combined with experience. However, now it is possible to gain reasonable estimates of what weather to expect by using forecasts based on a combination of statistics and mathematical modelling. Being able to model the potential variation in yields as a result of understanding the weather better enables companies that work

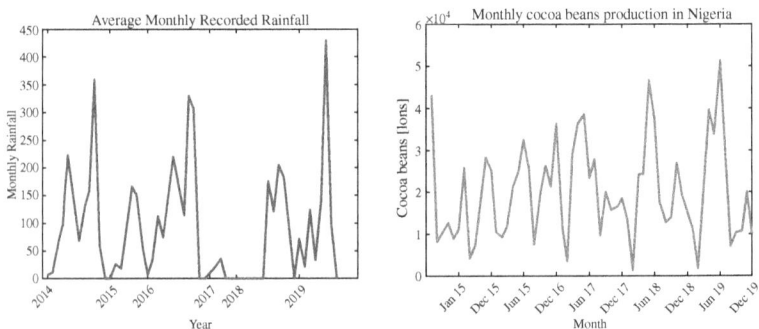

Figure 8.2   *Graphs of the rainfall (left) and the monthly cocoa harvest (right) in Nigeria showing the links. (Credit: With thanks to Mr Tosin Babasola and the Cocoa Research Institute of Nigeria)*

with the farmers and buy their crops to offer targeted advice to smallholder farmers on how to improve their yields, and help secure the development of sustainable farming communities. This work has the potential to make a real difference to the lives of farmers in Third World countries.

Mathematicians have also considered the growth of crops for the whole of the world. This is important if we are to grow enough to feed the world's entire population. The problem of how things grow, was studied in the classic text by D'Arcy Thompson [43] (now the subject of a major art exhibition in Edinburgh). If a relatively small amount of a crop is planted and allowed to grow year on year (during which it will be pollinated), then the rate of growth of the crop is proportional to the amount of the crop. We can express this as a *differential equation* of the form

$$dc/dt = a(t) \, c,$$

where the constant of proportionality $a(t)$ will depend upon factors such as the weather (rain fall, temperature and cloud cover) and the effects of irrigation and of pests. This equation works well if c is small, but as it gets larger, so more resource is needed to grow the crops, and the rate of growth of the crop slows down. Also, when c is large enough we will want to harvest it at a rate proportional to the amount of the crop. These effects are well captured by the so-called *logistic* equation introduced by Verhulst in 1838:

$$dc/dt = a(t) \, c \, (k - c) - b(t) \, c,$$

where k is an upper bound for the amount of crops and $b(t)$ is the harvesting rate. This equation can be solved to find the amount $c(t)$ of the crop, and is used to give a very useful prediction of its value if different harvesting strategies are employed and we have good predictions of the future weather.

An equation similar to the logistic equation was originally devised by Malthus to study population growth in human populations and we have already seen a discrete version of it in Chapter 5, when we considered the possibility of chaotic population growth. Such is the power of applied mathematics, that the logistic equation can be used in many other areas related to the supply of food. One of these is fishing, where now $c(t)$ gives the numbers of a fish species, and

$b(t)$ a strategy for catching the fish. Extra terms need to be added to allow for the movement of fish into and out of the fishing area, but the basic equation remains the same. See [5] for more details and other extensions to the model. Models for the populations of fish allow managements of fisheries to determine what level of fishing is possible to ensure that there is a sustainable fish population.

I should also point out, that whilst I have considered the role of maths in food production from fields or from the sea, there are other means of producing food which will become increasingly important as our need for food becomes greater. One of these is hydroponics in which food is grown on large water bags. Essential to the success of this process are the sciences of fluid mechanics and structural mechanics. Mathematics plays a key role in both. Mathematics is also hugely important in the correct application of fertilisers and insecticides and many other processes.

4. How to keep food fresh by freezing it, and then test it for freshness

Once food has been produced it needs to be delivered to where it can be eaten. Usually it is not possible to eat it at once, and it must be stored in such a way that it remains fresh. An early method of doing this was to salt it. However, a major breakthrough came with the widespread introduction of food refrigeration in the 19th and 20th Centuries. Now we take the refrigeration and freezing of food for granted. However, the revolution in cooling and freezing food was due in no small part to the discovery, and mathematical formulation, of the laws of thermodynamics by Lord Kelvin and others in the 19th Century. These allowed the development of efficient refrigeration devices based upon the expansion of gases.

When freezing food it is important to cool it at the correct rate to ensure that it is uniformly frozen. This requires us to predict both the temperature T of the food, and also the region of the food which is frozen. This is not an easy task. We will start by looking first at the simpler equation of how heat is conducted through a solid body.

The heat equation and its solution

The basic model for predicting the temperature of any solid material, including food, is called the heat equation. This equation was discovered by the French mathematician Joseph Fourier (1768–1830) when he considered how heat travelled down a bar of length L, which was initially at a uniform room temperature and was then was heated to a high temperature T_1 at one end and cooled to a low temperature of T_0 at the other. This situation is illustrated in Fig. 8.3, giving us Step 1 of the modelling process.

Fourier realised that the temperature $T(x,t)$ of the bar depended on time t and position x, and it changed throughout the bar by being conducted through the metal. Provided no part of the bar is frozen then (to shortcut Steps 2–3 of the modelling process) Fourier found that T satisfies a partial differential equation, the *heat equation*, which is given by:

$$\rho c \frac{\partial T}{\partial t} = k \frac{\partial^2 T}{\partial x^2}.$$

Here ρ is the density of the bar, c is its specific heat, and k is the thermal conductivity of the bar, which measures how well it is able to conduct heat.

The heat equation is an example of a partial differential equation. We have already met equations of this form when we looked at the equations governing weather and climate. (Indeed, the heat equation itself plays a very important role in our studies of climate as it tells us how heat is conducted through the atmosphere and the oceans.) Such partial differential equations are universal when we look at mathematical models of the real world. In general we cannot solve them mathematically and have to use a computer. It is not easy to solve

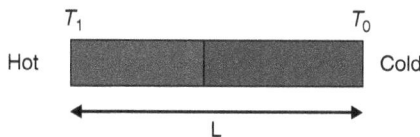

Figure 8.3   *A bar heated at one end and cooled at the other.*

the heat equation, but unlike the equations of climate it can be solved exactly. Indeed, Fourier came up with a brilliant way to solve it which is now called a *Fourier Series*. Using a Fourier series to solve a hard problem is rather like trying to build a house. Rather than building it in one go, it is much easier to build it brick by brick. The bricks that Fourier used to solve the heat equation were 'simple' solutions given by functions of the form

$$T(x,t) = e^{-\mu t}\left(a\cos(\alpha x) + b\sin(\alpha x)\right),$$

where a and b are arbitrary numbers, and μ and α are related by the formula:

$$\rho c\,\mu = k\alpha^2.$$

You can check this for yourselves by substituting this solution into the partial differential equation.

These solutions are very special, but Fourier made the astonishing intellectual leap of realising that the general solution of the heat equation could be found by combining an infinite number of these special solutions. If the bar has length L then Fourier considered a general solution of the form:

$$T(x,t) = \sum_{n=0}^{\infty} e^{-\frac{4k\pi^2 n^2 t}{\rho c L^2}}\left(a_n\cos\left(\frac{2\pi nx}{L}\right) + b_n\sin\left(\frac{2\pi nx}{L}\right)\right).$$

This rather fearsome expression is called a *Fourier Series*. If you look carefully you will see that it is made up of lots of special solutions of the heat equation multiplied by the numbers a_n and b_n. These numbers are called the *Fourier Coefficients*. In the case of $t = 0$ then the expression takes the form

$$T(x,0) = \sum_{n=0}^{\infty}\left(a_n\cos\left(\frac{2\pi nx}{L}\right) + b_n\sin\left(\frac{2\pi nx}{L}\right)\right).$$

Now, $T(x, 0)$ is the initial temperature of the bar, which we can assume is known. Fourier's genius was to realise that if you can work out the values of a_n and b_n for $T(x, 0)$ then the Fourier Series would allow you to work out the temperature $T(x, t)$ for all future times and locations.

Fourier's method works not only for the heat equation, but also for many other linear partial differential equations. It inspired a whole branch of mathematics now called *Fourier Analysis*. Not only is Fourier Analysis of vital importance in solving partial differential equations, it also has huge applications in most areas of physics and engineering. For example, if we want to reproduce a particular sound then we 'simply' have to work out the Fourier Coefficients of that sound. Since the discovery of the computer algorithm called the Fast Fourier Transform (FFT) then this can be done very quickly indeed. This process lies at the heart of the modern communications revolution.

One application of Fourier Analysis is a formula which allows us to calculate the Fourier Coefficients. In particular if the bar goes from $x = 0$ to $x = L$ we have that:

$$a_0 = \frac{1}{L}\int_0^L T(x,0)dx$$

and if $n > 0$ then

$$a_n = \frac{2}{L}\int_0^L T(x,0)\cos\left(\frac{2\pi nx}{L}\right)dx, \quad b_n = \frac{2}{L}\int_0^L T(x,0)\sin\left(\frac{2\pi nx}{L}\right)dx.$$

As an example of how we can use this, imagine a bar which is initially at a room temperature of 20°C so that $T(x,0) = 20$. The lefthand side of this bar is then suddenly raised to 100°C. Following the procedure above we can work out the temperature for all future times. The result is a rather messy infinite series, but we can easily sum it up on a computer. Figure 8.4 is an example of just such a calculation when we take: $L = \rho = c = k = 1$.

In Fig. 8.4 the left-most curve is the temperature immediately after the left-most point of the bar is raised to 100°C. You can see that all of the rest of the bar is at 20°C. The curves then show what happens as the time t increases to $t = 0.5$. You can see that the 'heated' part of the bar increases in length as the time increases. For long times the bar reaches a steady temperature $T_\infty(x)$ which is given by

$$T_\infty(x) = 20 + 80\,(1 - x).$$

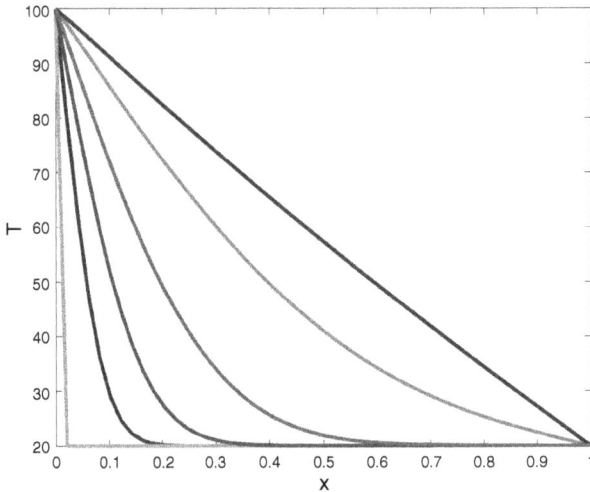

Figure 8.4 *The temperature of the bar for a set of increasing times. The temperature tends to an equilibrium of a straight line for large times with one end at 100° C and the other at 20° C.*

Icy times

If instead of a metal bar we think of a liquid such as water, or a food stuff which is mostly composed of water, then things become much more complicated, and we must extend our previous model. The reason for this is that if water reaches a temperature of 100°C then it boils and turns to steam (a gas). Similarly when water drops to 0°C then it freezes and turns to ice. This is called changing the state of the water, or a phase change. Turning water into steam requires extra energy, called the Latent heat. Similarly, is takes the addition of Latent heat to turn ice into water. The sum of the heat energy $Pc\,T$ needed to warm up the water to a temperature T (often called the Sensible Heat) together with the Latent heat at the phase change, is often called the Enthalpy $H(T)$. This idea was invented by the Dutch scientist Heike Onnes in 1909.

To find the temperature of the food stuff, and to allow for the change of state, we must extend our original model. In this extension the heat equation is replaced by the more complicated nonlinear *Enthalpy equation:*

$$\frac{\partial H}{\partial t} = k\frac{\partial^2 T}{\partial x^2}.$$

The Enthalpy equation is hard to solve by hand (the Fourier Series method won't work in this case because the equation is nonlinear). Nevertheless it is quite easy to solve on a computer. By solving this equation it is then possible to plan, and control, the freezing process for a wide range of different foodstuffs. In Fig. 8.5, we show just such a simulation. In this problem a column of water between $x = 0$ and $x = 1$, which is initially at room temperature, is kept at room temperature at $x = 0$, but the end at $x = 1$ is reduced to $-10°C$. It then starts to freeze from the right, with a region of ice growing from the right. In this figure you can see the temperature for an increasing series of times. The edge of the ice is visible as a 'corner' in each graph. As time increases, the water tends to an equilibrium in which the region $x > 0.7$ is all ice.

By solving the Enthalpy equation it is possible to predict with high accuracy how rapidly a foodstuff will freeze, and how much of it will remain frozen

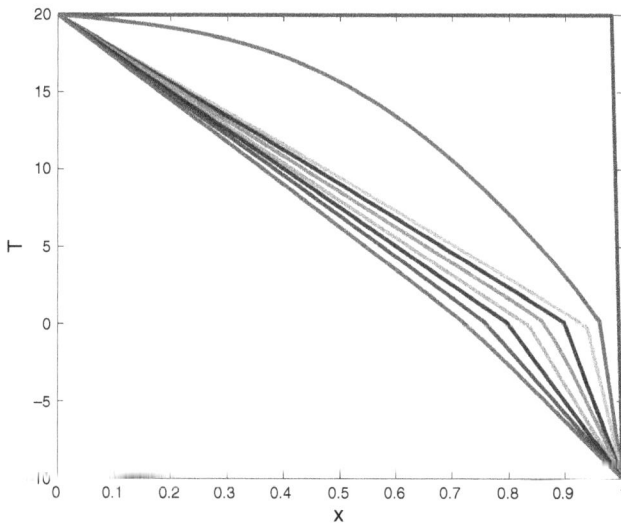

Figure 8.5 *A column of water in which the right-hand end is reduced in temperature to -10 degrees, and the left-hand maintained at room temperature. The water freezes from the right-hand side, finally forming a region of ice occupying the region $x > 0.7$.*

(as we saw in the example above). This knowledge is vital if we want to keep food frozen so that it remains fresh. Once food is frozen it is stored in refrigerated cabinets, rooms, trucks, buildings and warehouse, some of which can be as large as a football pitch. Such storage can bring its own problems. For example, what happens if the door of the warehouse is left open for too long. By calculating the transfer of the heat within the warehouse using the Enthalpy equation, it is possible to provide careful guidance for the safe time that this can happen without the frozen food deteriorating to a point where it is not safe to eat. Such procedures can potentially save huge amounts of food going to waste, and allow it to be stored and transported safely.

How fresh is a fish?

One of the more interesting problems that I have ever had to work on was that of finding out how fresh a fish is. When fish such as those shown in Fig. 8.6 are caught at sea they must be brought back to harbour and then sent on to (for example) supermarkets. Clearly such retailers need to know how fresh the fish is.

We often think that we can test the freshness of a foodstuff by its smell, but often food only starts to smell when it is far from being fresh. So, with fish, a

Figure 8.6 *Recently caught fish. But how fresh are they? (Credit: https://www.shutterstock.com/ image-photo/view-on-london-array-offshore-wind-1952137951)*

method for testing it had to be used which did not (only) rely on its smell. The method we came up with was to look at the *elasticity* of the skin, and the *viscosity* of the flesh beneath the skin. Both are closely related to the freshness. For example, our own skin becomes less elastic as we grow older. In order to test this method we produced a mechanical probe to test the elasticity of the fish skin. This probe bounced a small needle off the skin, and then monitored its response. By formulating a mathematical equation for the expected motion of the skin, and comparing this with the measurements of the probe, it was found possible to deduce successfully both the elasticity and the viscosity of the flesh, and hence the freshness of the fish. I will talk more about these equations in some detail in the next chapter about the mathematics of materials.

5. Why a bunch of mathematicians couldn't organise a piss up in a brewery

In the year 2005 the British Science Festival came to Dublin. At that time I had the honour of being the president of the Maths Section of the British Science Association. This august body had the responsibility of devising a mathematics programme for the event. One of our plans was to have a mathematics visit to the Guinness Breweries in Dublin. Obviously there are many reasons why we might want to visit a brewery, but why should mathematicians want to go there, and why should they want to go to Guinness? The answer to both of these questions lies in the person of William Gosset (1876–1937) pictured in Fig. 8.7.

William Gosset was the chief statistician at Guinness in the first part of the 20th Century. Guinness was in many ways ahead of its time in the production and quality control that it applied to its product (as well as the way that it was advertised). Gosset was employed in part to ensure that the Guinness stout was of a consistent quality. This was done by making careful measurements of a sample of the product and using these to assess both its general quality and its variability. This was, at the time, a difficult problem in statistics. To solve it Gosset devised a new statistical test to compare the measurements. This worked extremely well and made a very real difference to improving the quality of

Figure 8.7 *William Gosset (1876–1937), chief statistician for Guinness. (Credit: Wujaszek, https://upload.wikimedia.org/wikipedia/commons/4/42/William_Sealy_Gosset.jpg)*

Guinness Stout. Gosset felt it important to publish this test, but was reluctant to disclose his identity and that of his employer. Instead it was published in the journal *Biometrika*, in 1908, under the anonymous name of 'Student'. Ever since, this test has been known as *Student's t-test* and it plays a central role in testing and maintaining the quality of food and drink all over the world, as well as countless other products.

So, let's get back to the British Science Festival in 2005. Having decided to go to Guinness we set up a sub-committee to organise the trip during the science festival, in part to celebrate the invention there of the *t*-test, and also its contribution to modern statistics. Clearly such a trip should include a reception and a drink of a pint of Guinness. Unfortunately, through no one's fault, it wasn't possible in the end to do this. It was only after the event that we realised that we could then be accused of being unable to organise a piss up in a brewery.

6. I'm forever blowing bubbles

We have mentioned Guinness in the previous section. One of the key features of a pint of Guinness is the wonderful creamy foam head. This is in contrast

Figure 8.8 *A pint of Lager. Note the relatively small head at the top caused by bubbles of Carbon Dioxide. (Credit: https://www.shutterstock.com/image-photo/lager-on-tabletop-254078878)*

to the much smaller head that we find on a pint of bitter or lager beer and can be seen in Fig. 8.8. For the manufacturers of beer to get both types of head involves a lot of science and maths. The foam in the head in a pint of bitter is made of networks of bubbles of Carbon Dioxide separated by thin films of the beer itself, with surface tension giving the strength to the thin walls surrounding each bubble. The walls of these bubbles move as a result of surface tension with smaller bubbles moving faster as they have a higher curvature. This results in the smaller bubbles being 'eaten' by the larger ones in a process called Oswald ripening. Basically small bubbles shrink and large bubbles grow, leading to a coarse foam made up of large bubbles only. Eventually the liquid drains from the large bubbles and they pop, and the foam disappears.

Foams are very important in many applications, from food and drink, to fighting fires. A lot of effort has gone into devising mathematical models which explain their properties. The remarkable mathematician John von Neumann, who was (amongst many other achievements) responsible for the development of the modern electronic computer, devised an equation in 1952 which explained the patterns that we see in such cellular structures in two dimensions. In 2007 this was extended to three dimensions by a group of mathematicians in Princeton interested in the applications of maths to the foam on a pint of beer [45]. It's a hard life!

Another group of mathematicians, appropriately from the University of Limerick in Ireland, have made a study of the foam on a pint of Guinness as illustrated in Fig. 8.9. This is much creamier than the foam on a pint of bitter. The reason is that whilst the foam on bitter is made up of air bubbles, the foam

Figure 8.9 *A pint of Guinness Stout. Note the much thicker, and creamier, head caused by bubbles of Nitrogen. (Credit: https://www.shutterstock.com/image-photo/poltava-ukraine-march-22-pint-guinness-1106908061)*

on a pint of Guinness is made up of Nitrogen. This gas diffuses 100 times slower in air than Carbon Dioxide, meaning that the bubbles are smaller and the foam is much more stable and creamier. The Nitrogen needs to be introduced into the Guinness when it is poured. In a pub this is achieved by having a separate pipe, linked to a Nitrogen supply pumped in at the same time as the beer is served from the barrel. For many years Guinness in cans did not have a head. However, this problem was solved by the introduction of a 'widget' which is a Nitrogen container in the can, and which releases precisely the right amount of Nitrogen when the can is opened. This process must be very carefully controlled, and a lot of careful design work is required to make the widget work well. The whole process was analysed by using careful a mathematical model, derived (it appears!) by the whole of the applied mathematics department at Limerick and others. This mathematical model is described in the charmingly titled paper "The initiation of Guinness" [46]. Notably the same group has now done a complete analysis of the mathematics of making coffee.

7. Saving the penguins

Which came first, the chicken or the egg? The answer is of course the egg! Think about it in the context of the inheritance of parental characteristics. Maths isn't needed to solve that question. However, it is important both in the question of hatching an egg and also in helping the chickens, and other birds, to lay the eggs in the first place. The chicken industry is a huge part of the food industry, with an estimated number of 26 billion chickens in the world in 2020. For chickens to able to lay healthy eggs, they must be healthy. Keeping them healthy is a matter of giving them a good diet and also a safe environment. Statistical techniques are used extensively to determine good and healthy diets for all animals, and also to monitor how they respond to their environment, so that it is never too hot or too cold. (These tests are not unlike the clinical trials used by pharmaceutical companies to test drugs before they are released.) Now consider what happens when an egg is laid. If we want to breed more chickens these eggs need to be incubated, and it is most efficient (in the case of large numbers) to use an incubator to do this. Such incubators have to carefully regulate their heat and humidity. Sophisticated incubators also rotate

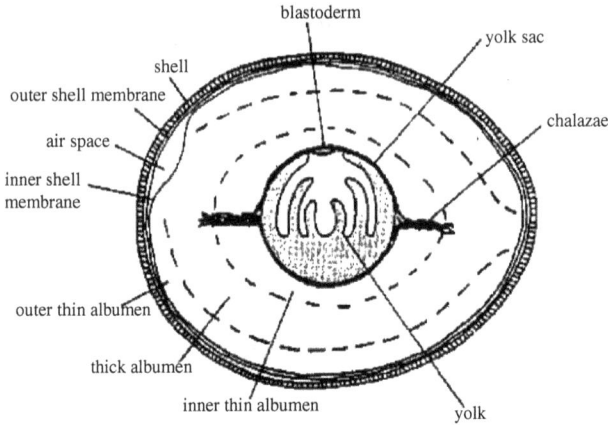

Figure 8.10 *A cross section through a typical penguin egg.*

the eggs during the incubation process and it is natural to ask the question of what is the best way to rotate the eggs. To answer this question we need to devise a mathematical model of the incubation process.

I was part of a team of mathematicians in the European Study Group with Industry in 2003 which was asked to do exactly this, not by a chicken farm, but by the penguin house at Bristol Zoo Gardens. The managers of the penguin house wanted to determine the optimal incubation process, and in particular the best way to rotate the eggs laid by a penguin. Unfortunately, the reason that we were asked to help was the fact that the zoo was finding that too many of the eggs in the incubator were dying. During a natural incubation process a (male or female penguin as both are involved) sits on the egg to keep it warm, and rotates it at the same time. In the case of natural incubation the egg is rotated about once every 20 minutes.

> One theory behind the need for such a rotation was that it was needed
> to ensure that the heat (from the penguin) was uniformly distributed.

We decided to test this theory by constructing a mathematical model of the heating process of a penguin egg as illustrated in Fig. 8.10 and Fig. 8.11.

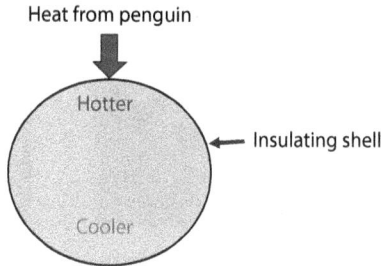

Figure 8.11 *A schematic of the heating of a penguin egg.*

As before we can go through the essential steps described in Chapter 1, with the basic situation illustrated in Fig. 8.11.

The basic physics is that the penguin sits on a part of the outside of the egg. The heat from the penguin is then transferred to the inside of the egg, which then starts to heat up. The rest of the egg shell acts as an insulator, preventing (or at least significantly slowing down) the heat in the inside of the egg from leaking out into the atmosphere. The material in inside of the egg then acts as a conductor of heat, in the same way as the metal bar did in the previous example.

In this case it is possible to use the same heat equation as was used to allow us to predict the flow of heat within the penguin egg. (We don't have to use the more sophisticated Enthalpy equation as the penguin egg never experiences a change of state. If it did then the penguin embryo would die.) The only difference is that the thermal conductivity, and the heat capacity of this material are different from that of a metal bar. These numbers are all known, or can be easily measured in an experiment.

Because the egg is three-dimensional (indeed approximately spherical) then the temperature T is a function of the three spatial variables (x,y,z) and also of time t. In the interior of the egg we then have the three-dimensional heat equation which is given by

$$\rho c \frac{\partial T}{\partial t} = k \left(\frac{\partial^2 T}{\partial x^2} + \frac{\partial^2 T}{\partial y^2} + \frac{\partial^2 T}{\partial z^2} \right).$$

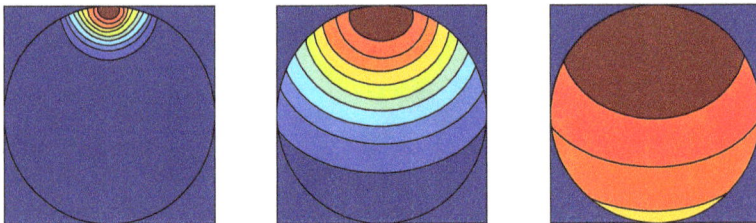

Figure 8.12   *The interior temperature of the egg (brown hottest and blue coolest) at (from left to right) times t = 1 minute, t = 5 minutes and t = 10 minutes.*

On the part of the egg in contact with the penguin we take $T = T_0$ = normal penguin temperature, and on the rest of the egg we take the normal derivative $\partial T/\partial n = 0$.

This equation is hard (though not impossible) to solve analytically (you use a three-dimensional version of a Fourier Series). However, it can be easily solved numerically on a computer by using a finite difference or a finite element method.

The pictures in Fig. 8.12 show the results of using such a numerical method to calculate the interior temperature of the egg at a set of increasing times from left to right given by 1 minute, 5 minutes, and 10 minutes. In these pictures the penguin is at the top, the brown areas are the hottest parts of the egg, and the blue areas are the coolest.

As you can see the heat from the penguin spreads through the egg and eventually warms it all up. For the particular case of a penguin egg we could calculate that this took about 10 minutes. So 10 minutes after sitting on the egg, it has reached a nearly uniform temperature, and at 20 minutes the temperature is a completely uniform, penguin bottom temperature.

In other words, there is no need to rotate the egg at 20 minutes to give it a uniform temperature, as it is already at a uniform temperature.

> *Thus the simple mathematical model was useful in dismissing an incorrect theory for why the penguin rotates the egg.*

Further studies of why the egg had to be rotated needed a much more sophisticated mathematical model. This model looked at the different types of fluid inside the egg, the difference between the yolk and the albumen, and the various chemical and biological process that go on inside the egg. These models showed that the egg needed to be rotated to both redistribute the nutrients inside it and also to get rid of the waste products. Guided by this a better rotation strategy was devised. This strategy led to an increase in the number of eggs that hatched into live penguins. More information not only about the penguin egg problem, but also more about the process of mathematical modelling can be found in Howison's textbook [47].

8. Cooking food in a microwave cooker

Before we eat food it is usual to cook it. This is one huge difference between humans and other animals. In fact evidence for cooking food goes back a long way, including evidence that our evolutionary predecessors Homo Erectus were cooking food over 400,000 years ago. Of course cooking then was done over an open fire. Nowadays it is more likely that we will use some kind of cooker. Until recently, most cookers were based on heating the food directly, such as in a fan oven, which cooks the food from the outside by conduction of heat into the food. This process cooks the food uniformly, but can be slow. For example, it takes about an hour to bake a potato in a typical fan oven. Many of us now seek a more convenient and faster means of cooking, and this has led to the rise of the use of the microwave cooker. The microwave cooker uses a technology which goes back to the war and the invention of radar. In order to get a high resolution (particularly for airborne radar), radar systems needed to use radio waves, called Microwaves, with a wavelength of the order of just a few centimetres. However, in 1940 there was no stable means of producing radio signals at this wavelength in sufficient power to be effective.

Figure 8.13 An early cavity magnetron. These were used in radar sets World War II. Magnetrons are now used to produce the microwaves used in a microwave cooker. (Credit: Geni, https://commons. wikimedia.org/wiki/File:Randall_and_Boot_cavity_magnetron_Science_museum.JPG)

Fortunately the University of Birmingham came to the rescue in the shape of the physics department led by Professor Oliphant. Working in this department were Randall and Boot who invented the first high power *cavity magnetron* pictured in Fig. 8.13. This device used a high magnetic field to spin an electron beam and to cause it to resonate in a specially designed cavity. The result was a method of producing high power radio waves of kilowatt power and at Centimetre wavelengths. The magnetron revolutionised radar and was subsequently used in all airborne interception radars and also in the H_2S air to surface radar used in Lancaster bombers, as well as in the radars used against U-Boats. It was truly a war-winning invention (and was part of a package of secrets taken to the USA by the scientific Civil Servant Sir Henry Tizard as part of the process of persuading them to join in the war). The Americans rapidly developed the technology for producing magnetrons in large numbers. One of the scientists who did this was Percy Spencer, and he is credited with the invention of the microwave cooker. Legend has it that he did this after noticing that his candy melted when exposed to high power microwave radiation, and he realised that the same radiation could be used to rapidly cook food.

Now microwave ovens, all powered by magnetrons, are very widely used in a domestic environment. We are all used to opening up a microwave cooker, putting in some food, pressing the button, and ping, five minutes later the food is cooked. Usually these ovens have a turntable, and you can occupy yourself in those five minutes by watching the food go round. Mathematical modelling is very useful not only in understanding this process, but in helping to devise better and safer microwave cookers. This was a task that I was set by the Chipping Campden Food Research Association (CCFRA).

As always we will start Step 1 of the mathematical modelling process by considering the basic process of what goes on in a microwave cooker. A commonly held view is that a microwave cooker cooks food from the inside out. However, this is not really true, as we shall see. A typical such cooker (indeed the one that we used in our experiments at CCFRA) is illustrated in Fig. 8.14 (together with a set of temperature sensing probes).

The magnetron generates the microwaves for the cooker. Typically these magnetrons have a total power of around 800 W. The microwaves produced by the magnetron are rapidly oscillating electromagnetic fields with a very high frequency of around 2.45 GHz and a wavelength of 12.24 cm. The microwaves then enter the large cavity in the middle of the cooker where they set up standing wave patterns in the electromagnetic field in the cavity. These standing wave patterns give points with a strong microwave field alternating with points with a low field. To show these points you are warmly recommended to try out the

Figure 8.14   *A typical microwave cooker. This was the cooker that we used for our experiments. You can see the tub in the middle where the food was placed, and the temperature sensing probes.*

following experiment. Take out the turntable and place several marshmallows in the cavity. Then turn on the microwave oven for a short time. You will find that some of the marshmallows have melted and others have not. The melted ones are at the locations of the anti-nodes of the microwave field, which are the points where it is strongest. As an added bonus you can find the half wavelength $\lambda/2$ of the field by measuring the distance between the melted marshmallows. Typically, as we said above if the frequency is 2.45 GHz then $\lambda = 12.24$ cm. You can now check this directly using the power of the marshmallow!

The nodes of the microwave field are where it is weakest. If the food is placed in one of these then it would hardly be cooked (this would be a *cold spot* in the food). The reason that the food is placed onto a rotating turntable is to make sure that it is constantly moving, so that no part of it is always at a node of the field.

During the cooking process, the microwaves from the cavity enter the food and start to heat it up. The way this happens is that the very high frequency microwaves cause the water molecules in the food to oscillate at the same frequency. They then rub against each other and warm up the rest of the food by friction. This transfers energy from the microwaves to the food, heating the food up in the process. As they do so, the microwaves rapidly lose their strength. This process works very well for water, and hardly at all for ice. So microwave cookers are great for heating up moist foods, but are poor at defrosting frozen foods.

We will now do some mathematics to explain how the field reduces as the microwaves penetrate the food. Key to this is what is called *Lambert's Law*. This says that if E_0 is the average strength of the microwave field on the *surface* of the food, then at a distance x into the food, the average strength $E(x)$ of the field inside the food is given by the formula

$$E(x) = E_0\, e^{-x/d}.$$

Here d is called the *penetration depth* of the microwaves. The value of d depends on the type of the food, how *moist* it is, and also its salt content. For a typical food, such as a potato (which is mainly starch and water) the penetration depth is between one and two centimetres. What this means is that the microwaves cannot penetrate much more than two centimetres into the food.

Now, the heating *power* of the microwave field is proportional to the square of the amplitude of the microwave field which is given by $|E(x)|^2$. We saw a very similar expression for power in terms of voltage in the last chapter. This power is *negligible* for fields at depths x much greater than the penetration depth. Therefore, the microwaves produce heat *within* the food up to, but not much deeper than, the penetration depth. Heat is then mainly lost on the *surface* of the food, mostly by convection through the air but also due to radiation. Because of this the temperature of food heated in a microwave oven initially *rises* as you go in from the surface, and then starts to fall after you have gone in further than the penetration depth. So the best way to describe microwave cooking is that the food is cooked from about 1 cm underneath its surface. There is an important consequence to this. If the food is much more than a centimetre or two in depth, then the inside of it may receive *very little* direct heating from the microwave field at all. Therefore, if the food is only placed into the microwave cooker for a short time, the inside does not get heated to a high temperature, and bacteria in the food may not get killed. This means that the food may then be unsafe to eat.

Now let us think about how we will model the microwave cooking process. We will use the same heat equation that Fourier devised to find the temperature of the heated bar, but with the very important difference that we will now include a term which models the heat generated by the microwaves. The power $P(x)$ that the microwaves supply to heat the food at the point x is then given by

$$P(x) = \mu \, E(x)^2,$$

where the value of μ depends upon the type of food and how much water it contains. (This number has to be found in advance by experiment.) Using the expression for $E(x)$ given by Lambert's Law we can deduce that

$$P(x) = \mu \, E_0^{\,2} \, e^{-2x/d}.$$

Combining this with Fourier's heat equation we arrive at our model for the temperature $T(x,t)$ which is given by the following partial differential equation:

$$c \rho \, T_t = k \, T_{xx} + \mu \, E_0^{\,2} \, e^{-2x/d}.$$

To use this equation to predict the food temperature we must estimate the average electrical field E_0 on the surface of the food, which we can calculate once we know the power of the magnetron. We also need to know the way that heat is lost on the surface of the food, and this will mean knowing what sort of packaging the food is in. We also need to account for the *shape* of the food. Typically a microwaved meal comes in a rectangular container, so it is a fairly uniform shape. The formula above is then applied to each of the sides of the food in turn.

Figure 8.15 taken from the Budd and Hill paper [48] show the results of solving this equation to find the temperature of a rectangular plastic container filled with mashed potato, which is mostly made up of moist starch. This simulation is achieved by solving partial differential equations, which describe how the potato is heated. In fact, to allow for the possibility that the moisture in the starch may reach the boiling point of $100°C$ and turn to vapour (which then escapes through the food) the heat equation above must be extended to a form of the enthalpy equation which looks like:

$$H_t = k \, T_{xx} + \mu \, E_0^{\,2} \, e^{-2x/d}.$$

The figures are all created from solving the enthalpy equation above on a computer with the correct values for all of the terms. (The calculations were done by my PhD student Andrew Hill.) The top shows the results of heating for one minute and the bottom of heating for five minutes. In these figures blue is cold and brown is warm. In this simulation we show the temperature through a cross section of the rectangular container. This is mainly exposed to the microwaves from the top and the bottom, and it has thermally insulating

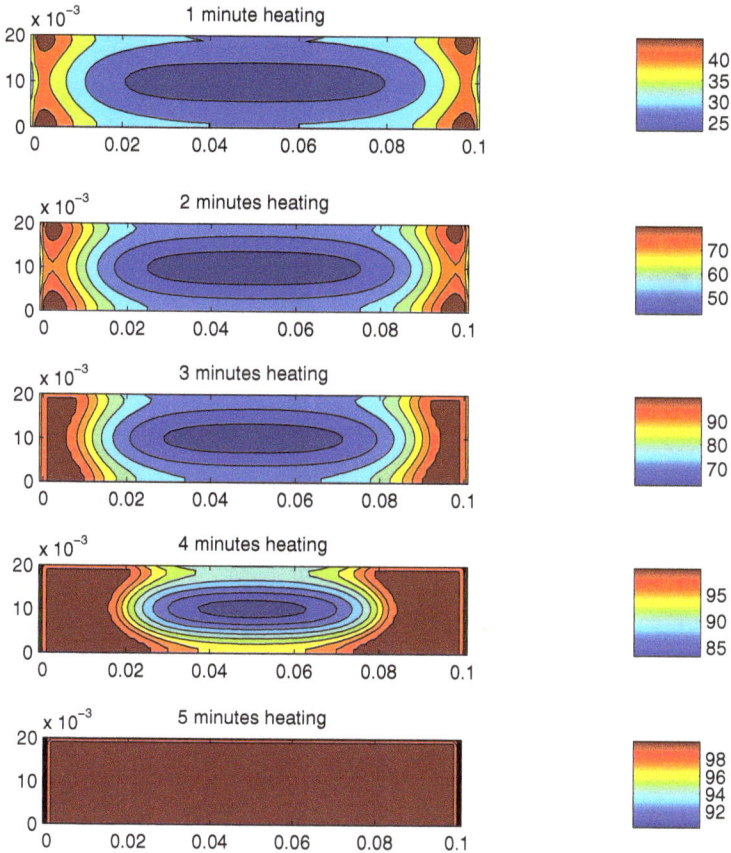

Figure 8.15 *A cross section through a rectangular container containing moist starch heated in a microwave cooker. The plots show the temperature of the starch modelled by solving the enthalpy equation with the addition of the source term for the power from the microwaves. See [9] for more details.*

walls. Heat is mainly lost to the air at the top and the bottom. The dimensions of the rectangular container are all expressed in metres.

You can see from these figures that in the first few minutes the walls of the food are hot. We would feel this if we tried to lift the food out of the cooker. However, this would be deceptive, as the inside of the food is still quite cool. Only after five minutes is the food uniformly hot, and this is due

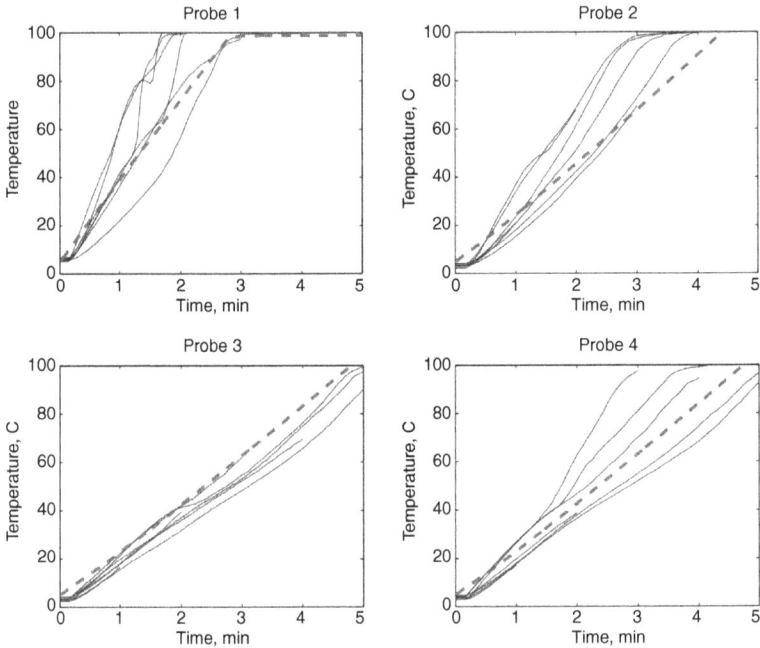

Figure 8.16 *A series of temperature measurements of the microwave heating experiment taken from different probes within the moist starch. The experimental measurements are shown in solid lines, and the predictions of the model in the dashed line. See [48] for more details.*

mainly to conductance of the temperature within the food evening the temperature up.

Of course, as part of the modelling process that I have described, we needed to check the results of our mathematical model against experiment. To do this we put temperature probes into the food at a series of four points, with the first probe nearest the outside and the fourth nearest to the centre. You can see these if you look carefully at the microwave cooker in Fig. 8.14. We then ran five different trials of heating the moist starch in the microwave cooker, and measured the temperature each time. The results of doing this are shown in Fig. 8.16 in solid lines, with the predictions of the model in the dashed line. You can see that these are in good agreement, including the phase change when the temperature reaches 100 degrees. See [48] for more details both of the model and of the experiments that were made to test it.

The conclusion from the simple model simulation that we might draw is that the best way to be sure that your food is warm all over when you use a microwave cooker to cook it for a long time, and then wait until the middle warms up by heat conduction. However, such long cooking times can damage, and even burn, the outside of the food which is most heated by the microwaves. An alternative method is to heat the food for a while, to then stir it to give it a more uniform temperature, and to then heat it up again. If you read the packet of any microwave cookable meal this is exactly what it says that you should do. Now we can see from our mathematical model why this is sound advice.

More sophisticated mathematical equations can be used to simulate more of the processes involved in microwave cooking and can thus be used to design better, more efficient and safer cookers. This brings great benefits to us all.

9. Packing and distributing

Once we have produced food we need to pack it and distribute it (see Fig. 8.17). Both of these processes involve the mathematics of optimisation and scheduling. Food needs to be transported around in such a way that it arrives at the correct place, at the correct time, and is fresh when it arrives. This introduces huge

Figure 8.17 *A typical freezer lorry used to transport food. (Credit: https://www.shutterstock.com/image-photo/watford-uk-june-13-white-refrigerated-1758254117)*

logistical challenges, made worse by the fact that different foods have different shapes, weights and times of delivery. We often forget how much planning is needed to (for example) deliver fresh strawberries to our plate in the middle of winter. These problems are very hard to solve. A classical example being the *travelling salesman problem,* which aims to find the optimal route for a salesman to deliver his goods. Another example is the *knapsack problem* which tries to find the best way to fit a set of differently shaped objects into a knapsack, with direct application to the problem of fitting food into a freezer lorry (illustrated) or a transport plane. Only relatively recently have efficient (probabilistic based) algorithms been developed to provide an answer. These algorithms are now making a huge difference to the way that goods are transported all over the world. To solve this problem uses the mathematical ideas of operational research, and the great subject of mathematical algorithms.

10. In conclusion: Three mathematicians go into a pub

I will finish this chapter with a bad joke about mathematicians and drinking. You have to concentrate a bit to get the joke.

Three mathematicians go into a pub and the bar tender asks, 'Does anyone want a lager?

The first mathematician pauses for thought, and then says, 'I don't know'.

The second mathematician likewise says, 'I don't know'.

Finally the third mathematician says, 'No!'

So the bar tender asks 'Does everyone want a bitter then?'

The first mathematician pauses for thought, and then again says, 'I don't know'.

The second mathematician likewise says, 'I don't know'.

Finally the third mathematician says, 'Yes!'

So they all have a bitter!

Now, think about the mathematical logic contained within this joke as you ponder the rest of the points in this chapter and consider the very real contribution that mathematics makes to feeding the whole of the world.

9
Material mathematics

1. Introduction

Materials dominate our lives. The clothes we wear, the tools we use, the cars that we drive, the aeroplanes we fly in, and the houses we live in, are all made up of materials. See the book [49] for an excellent summary of the many materials we make use of in our lives. Modern electronics would not be possible without the development of semiconducting materials, display screens rely on recent advances in liquid crystals, and the development of the jet engine happened at the same time as the invention of modern heat resistant alloys. Our history has even been defined in terms of the materials that we use, from the stone age, to the bronze age to the iron age, and now the computer age. Maybe the age we live in should now be called the *composite age*. In the past materials were either natural, such as wood and stone, or evolved through many years of experience, such as glass or paper. The properties of these materials were either beyond our control, or were obtained through experiment or even good luck. However, with our modern understanding of physics and chemistry, this has all changed. In particular we can now design meta-materials with given mechanical, electrical, optical, thermal and other properties. Indeed, modern materials are often composites of different materials with very different properties arranged in a complex manner. The resulting behaviour of these materials often *emerges* from the way that these properties interact. These are called meta-materials, which are purpose designed to have special types of behaviour. The tool that we have to make sense of this complexity, and to design the materials that we need for the modern world, is mathematics.

In this chapter we will start by looking at how materials bend and deform under pressure. This is very important in engineering. For example, it is essential to know how strong the materials in a bridge or a building are, if what they build is going to stand up to the stresses and strains that it will encounter in its working life. We will start by looking at beams and we will see how mathematical models can help us to predict the behaviour of that most ancient material, rock.

> *Warning! The mathematics in this chapter is pretty heavy duty and involves some rather advanced concepts. Materials are hard to analyse! In the next chapter the maths (of space) won't be anything like as difficult.*

We will then go on a journey through a series of materials, and crystals, including the possibility of making the invisibility cloak worn by *Harry Potter*. Finally, as a bit of light relief, we will have some fun with the modern mathematics of folding paper and origami.

2. How materials bend

Perhaps the earliest materials our ancestors made use of were wood and stone. Wooden tools don't survive for long, but there are examples of stone tools which are believed to be over one million years old. Rock is itself a very interesting material, and its use for tools and then later for building, has shaped human civilisation. Indeed we might say that rock is our longest lasting material, in some cases as old as the Earth itself. When we look at a rock we see something very solid. Indeed this is its main property as a material. However, appearances can be deceiving, and rock has many other properties. Some of these are very obvious to us all. Anyone who has been in an earthquake will have experienced the fact that rock can behave elastically. An elastic medium is one which can be distorted a small amount by a (suitably large) force, with a displacement proportional to the force. Crucially, when the force is removed, an elastic medium returns to its original position. An example of an elastic medium is a wire or a spring. The relationship between the extension of a spring and the force on it was, splendidly, first discovered by a predecessor of mine as Gresham Professor of Geometry. The amazing polymath Robert Hooke (1635–1704). His portrait (or at least a picture of what we think that he looked like) hangs in Gresham College in Barnard's Inn Hall and is illustrated in Fig. 9.1. In this picture we can see that he is holding a spring in his left hand. It was his work on the elastic properties of springs, summarised by Hooke's law, which lies at the heart of our understanding of how materials bend.

Figure 9.1 *The modern portrait of Robert Hooke which hangs in Gresham College, London. You can see that he is holding a spring in his left hand. (Credit: Rita Greer, https://en.wikipedia.org/wiki/ File:18_Memorial_at_Gresham_College.JPG)*

One feature of an elastic material is that it admits elastic waves of expansion and compression. In a spring we can see this by sending a pulse along its length. In just the same way, rocks transmit waves and we see these in the shock waves associated with earthquakes. As well as being destructive these wave scan also be very useful. Indeed much of our knowledge of the Earth's interior comes from studying the way in which these waves are reflected, refracted and a diffracted through the rock. By solving an inverse problem (this is a problem in which we try to find the cause of a measured effect) we can then find out the properties of the Earth's core. From a mathematical point of view this involves solving a *wave equation*. A very similar idea is used by oil companies to prospect for oil. To do this they detonate a small explosion (usually with compressed air) and measure the strength of the reflected waves. This allows them to detect oil deposits deep under the seabed. As we saw in Chapter 3 this can be bad if you happen to be a passing whale.

The above observations show that over short timescales (which to a rock is anything up to about ten thousand years), a rock will behave elastically. However, over longer timescales rock can behave like a different type of material. We are used to seeing rocks as cold, hard and brittle. However, during the course of their existence, rocks can be both hot, under extreme pressure and (over sufficiently long periods) can flow almost like a liquid. This sort of behaviour is called *viscous* and unlike an elastic material, in which displacement is proportional to force, a viscous material has a velocity, which is proportional to the applied force. Understanding all of these aspects of a rock as a material allows mathematics to be applied to the fascinating question of understanding how rock layers buckle under the application of a large force and over long periods. The reason that this is important is that under the effects of continental drift, rock layers are subjected to huge pressures from moving plates, and these forces deform the rock layers to produce mountains. Mathematical models can thus be used to predict how rocks fold and thus how mountains form.

We will now see if we can model how a rock layer deforms when it is subjected to a compression from a moving tectonic plate. We will follow our tried and tested modeling process to do this, starting with thinking about the basic processes involved. Our model will then progress through three stages of increasing sophistication.

A typical rock layer being deformed by a pressure P is illustrated in Fig. 9.2.

We can think of this layer as being like a very thick sheet of cardboard which is being pushed by a large continent which exerts a pressure P on the sheet. Try this yourself with a thin sheet of card (paper won't work as it is far too

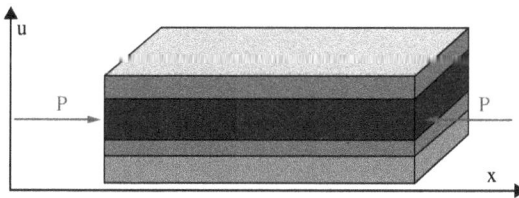

Figure 9.2 *Layers of sedimentary rock being compressed with a pressure P.*

floppy). The card has a stiffness which means that it resists being bent. This is why a cardboard box retains its shape. However, if you push the card hard enough (but not too hard) then it does bend. Release it and it should return to its original shape. This is called an *elastic* deformation If you push it really hard then the cardboard will crease and will not return to its original shape. This is called a *plastic* deformation. We can see both types of deformation occurring in cardboard, and both occurring in rock. (Roughly speaking what is happening at a molecular level is that in an elastic deformation the bonds between the molecules in the material are stretched, whereas in a plastic deformation they get broken.)

We will now follow our usual practice of building up a model for how the layer deflects by starting with a simple situation and making it progressively more complicated.

Case 1. Bending a beam

We start by thinking of the deflection of a beam of length L which occupies the space $0 < x < L$, and has a vertical deflection given by $u(x)$. Imagine that this beam has a *vertical load* per unit length given by $q(x)$. Working out how this beam will deflect is a problem with obvious engineering applications when it comes to designing buildings. Both Leonardo da Vinci and Galileo studied it, but a (partial) solution had to wait till around 1750 and was due to Leonhard Euler and Daniel Bernoulli. They derived the differential equation (often called the *bending equation*) for the change of $u(x)$ in space, which is given by:

$$EI \frac{d^4 u}{dx^4} = q(x).$$

Here I is the cross-sectional area of the beam, and E is the 'Youngs Modulus' which depends on the material of the beam. The larger that E is the stiffer the material. So E is very small for paper, moderate for cardboard, and large for rock or steel.

What is special about this equation is that it is *fourth order* because it has a term which involves four differentials of *u* with respect to *x*. This is unlike the earlier equation that we saw for the pendulum which was an example of a second-order differential equation. The higher the order of the equation the more complicated its solutions tend to be (and the harder it is to solve).

The bending equation above was used by Eifel in the 19th Century to design the Eifel Tower in Paris! It is now very widely used by structural engineers to design many different types of building, made from widely different types of material. It can also be used by biologists to study bones, and even the workings of the human ear.

To see how this equation works, imagine that the beam is hanging under its own weight only. In this case $q(x) = -\rho g$ is a constant, where ρ is the density of the beam, and g is the acceleration due to gravity. It is negative because the weight of the beam acts to deflect it downwards. We then have

$$\frac{d^4 u}{dx^4} = -\frac{\rho g}{EI}.$$

We can integrate this equation up four times to give

$$u(x) = -\frac{\rho g}{EI}\frac{x^4}{24} + ax^3 + bx^2 + cx + d.$$

Here the values of a, b, c and d are unknown at this stage of the calculation. To find a unique solution we need four extra boundary conditions which describe how the beam is supported at each of its ends. This is quite a complex subject, and the answers depend upon whether the beam is free to move, firmly clamped, or supported on a hinge or a pin joint. One example is a *cantilevered beam* which is clamped at the end at $x = 0$ and is free at the end $x = L$. In this case the boundary conditions are:

$$u(0) = u_x(0) = 0 \quad \text{and} \quad u_{xx}(L) = u_{xxx}(L) = 0.$$

If we take a case in which $L = 1$ and $\rho g/24EI = 1$, and apply these boundary conditions, then:

$$u(x) = -x^4 + 4x^3 - 6x^2.$$

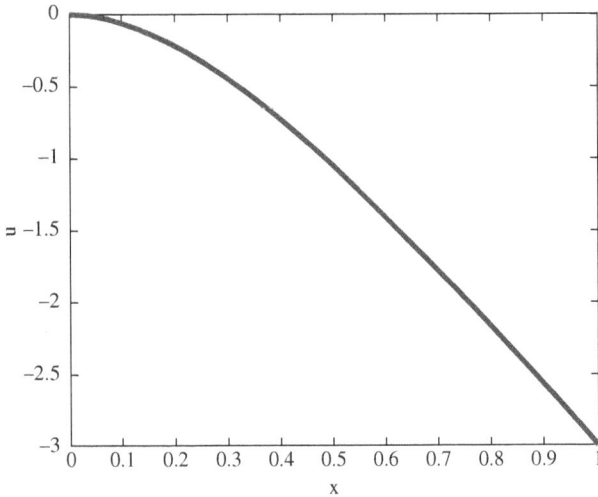

Figure 9.3 *The deflection of a stiff beam under its own weight. In this figure the beam is clamped at the left end.*

The deflection of the beam predicted by this model is illustrated in Fig. 9.3.

Case 2. Compressing cardboard

Now, we go back to our experiment in which we try to compress a layer cardboard. In this case we supply a horizontal pressure force P. If we now extend our model to include this then we get a more complex fourth-order differential equation given by

$$EI\frac{d^4u}{dx^4} + P\frac{d^2u}{dx^2} = q(x).$$

A special case of this problem is given by taking $q(x) = 0$ (which would correspond to having very light cardboard). We can then integrate the above equation twice to give

$$EI\frac{d^2u}{dx^2} + Pu = ax + b.$$

Here a and b are again arbitrary constants. If the boundary conditions at both ends of the cardboard are the same then we can take $a = 0$. We now have a second-order differential equation which is in every way the same as the simplified equation for the pendulum that we looked at in Chapter 1. This is the true power of applied mathematics. Quite different problems often lead to the same equation, and then the solution of one problem tells us the solution of the other. Using the results given in Chapter 1 we can then write down the general solution which is given by

$$u(x) = \frac{b}{P} + A\cos(\omega x) + B\sin(\omega x), \quad \omega^2 = P/EI.$$

Here A and B are again unknown constants, which we need to work out from the boundary conditions at the end of the cardboard. There is a big difference between this solution and the earlier one for the beam. That is, we have terms involving sine and cosine. This means that the deflection $u(x)$ of the cardboard can have oscillations in x. But this is very reasonable. Imagine pushing a carpet. It is easy to have a ripple in the carpet, which gets worse the more we push it. Our model for the cardboard has predicted exactly this effect. A typical deflection of the cardboard layer is shown in Fig. 9.4. See if you can compress your layer of cardboard to look like this.

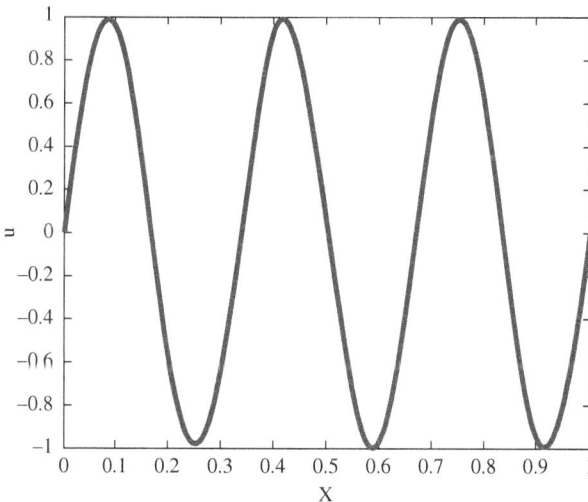

Figure 9.4 *Ripples in a compressed layer of cardboard.*

Case 3. Compressing and bending rock

Now we look at rock. In some ways a rock layer compressed from the side is like a cardboard layer, Whilst they might look very different, they are both stiff, and deform elastically in much the same way (just over very different time scales). However, there is one big difference. A rock layer is usually surrounded by other layers of rock. This is because sedimentary rocks are typically laid down at the bottom of the sea, with one layer formed on top of another over millions of years. When these are compressed by the forces of continental drift, then all of the rock layers buckle together. The effect of this is that the force $q(x)$ in the expressions above now represents the force on the rock layer of all of the layers above and below it. To imagine what this is like simply go to bed. When you do this you will probably lie on a mattress. The mattress deforms to support you and to give you a good night's sleep. To a rough approximation the rock surrounding the rock layer acts in a similar way to the mattress and you act like the layer itself. By lying on the mattress you deform it, in just the same way as the rock layer deforms the rock around it, and in the same way the surrounding rock pushes back on the rack later, in the same way as the mattress pushes back on you. The force $q(x)$ then represents exactly how much the 'mattress' of rock is pushing on the layer of rock. Typically the more the rock layer deforms the greater this force is. It is difficult to give an exact description of this force as it depends a great deal on the type of rock. However, at this stage we can use a device often used in modelling. If we don't know something exactly, then we give a reasonable expression for it which involves some unknown parameters. We can work out suitable values for these parameters later when we fit the model to the data (this process is called calibration). A typical such expression is to take

$$q(x) = -a\,u(x) - b\,u(x)^2 - c\,u(x)^3,$$

where a, b and c are the parameters to be determined. Combining all of this together we finally arrive at our model for rock bending. This is given by the following fourth-order differential equation:

$$EI \frac{d^4u}{dx^4} + P \frac{d^2u}{dx^2} + au + bu^2 + cu^3 = 0.$$

Having arrived at our mathematical model we will now try to solve it to find the shape of the rock. If you remember the rather similar looking equation for the pendulum, then we were unable to solve this in general because it was *nonlinear*. The same situation applies here. The nasty u^2 and u^3 terms put this equation beyond the reach of an analytical solution. But before we head for a computer to solve it for us, we recall that when we looked at the pendulum problem we could gain a lot of insight by considering small oscillations which satisfied a *linear* equation. In the case of the rock layer this linear equation is given by the following expression which is well known in the geo-sciences:

$$EI \frac{d^4u}{dx^4} + P \frac{d^2u}{dx^2} + au = 0.$$

This equation looks rather formidable. But we can solve it using the complex numbers which proved so useful in Chapter 7. To do this we consider a solution $u(x)$ which takes the form $u(x) = e^{\lambda x}$. Substituting this into the fourth-order differential equation leads to the expression

$$(EI\lambda^4 + P\lambda^2 + a)e^{\lambda x} = 0.$$

If we divide this formula through by $e^{\lambda x}$ we find that λ satisfies the *quartic equation*

$$EI\lambda^4 + P\lambda^2 + a = 0.$$

It is usually hard to solve quartic equations (there is a general formula but it is fearsomely complicated), but not in this case. This is because if we set $\mu = \lambda^2$ then we get the following *quadratic* equation for μ

$$EI\mu^2 + P\mu + a = 0.$$

We can solve this quadratic equation by using the formula given in Chapter 5, and there are two roots given by:

$$\mu = \frac{-P \pm \sqrt{P^2 - 4aEI}}{2EI}.$$

Now one of two things can happen. If there is a lot of pressure acting (as we might expect if the pressure comes from continental drift) then $P^2 > 4aEI$ and *both* roots of the quadratic equation are real and negative. In contrast if the pressure is much lower, with $P^2 > 4aEI$ then the roots are complex and are given by:

$$\mu = \frac{-P \pm i\sqrt{4aEI - P^2}}{2EI}.$$

Our modelling assumption is that the rock is being compressed by continental drift which leads to a very large (indeed almost infinite) pressure acting on the rock. In this case μ takes two negative real values. Now we remember that $\mu = \lambda^2$, and we must take the square root of these two negative values to give four possible values of λ. All of these values which will be purely imaginary (multiples of the imaginary number i only). This means that there are values α and β so that

$$\lambda = \pm i\alpha \quad \text{or} \quad \lambda = \pm i\beta.$$

Now, we also remember that $u(x) = e^{\lambda x}$. Substituting the values for λ above into this equation for $u(x)$ and using Euler's formulas for $\cos(x)$ and $\sin(x)$ which we met in Chapter 5, then we arrive at the following formula for the way that rocks bend:

$$\boxed{u(x) = A\cos(\alpha x) + B\sin(\alpha x) + C\cos(\beta x) + D\sin(\beta x).}$$

This is called a *quasi-periodic solution*. It is the sum of periodic functions of *different frequencies*. We saw a similar quasi-periodic function when we looked at the solution of the forced pendulum. The graph of the predicted shape of the folded rock is shown in Fig. 9.5, and certainly resembles the patterns seen in real rock layers.

Pictures of rocks folding in this way are often found in textbooks and in real life. An example of such is given in Fig. 9.6, taken from a photo I took of small-scale folds in rocks in the cliffs outside the seaside town of Bude on the North West coast of Cornwall. (Rumour has it that my surname originated in Bude.)

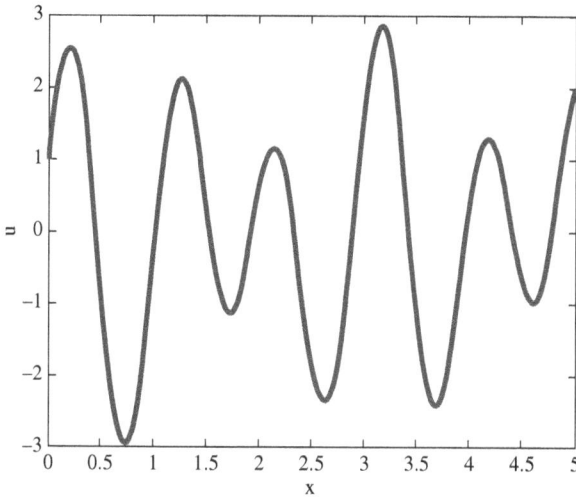

Figure 9.5 *The quasi-periodic shape of rock deformed under a very high pressure.*

Figure 9.6 *Quasi-periodic folds in rocks at Bude in Cornwall.*

However, quasi-periodic folding it is not the only shape that you find when rocks fold. The reason for this is the action of the extra nonlinear terms in the model that we are looking at. These can lead to much more exotic folding patterns.

To check this we followed Step 6 in the modelling approach of Chapter 1, and compared the results of our model predictions against an experiment. Now, we don't have many sedimentary rocks at Bath, and certainly don't have a device strong enough to compress a rock. However, thanks to a university reorganisation, and a change in the name of our department (it happens!) we did have a lot of redundant visiting cards. Stacking them in layers gave us a cardboard model of sedimentary rock layers. These we could then compress. This allowed us both to test the model above, and also to see how the layers of rock (sorry, cardboard) interacted with each other. My very good friend Giles Hunt, Professor of Structural Engineering at the University of Bath, had a press strong enough to compress these layers and was very keen to see the results. If you look at Fig. 9.7, then you can see the results of our efforts. They key thing to see is that we don't get the simple sine-waves of rock folding patterns often illustrated in textbooks. Instead we get folds which are like zig-zags. They have patterns which are a combinations of straight sections joined together by sharp bends. Geologists call these folding patterns kink bands and chevrons.

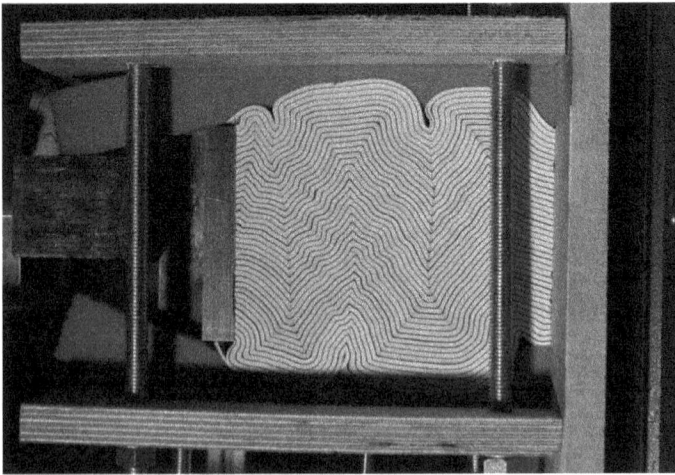

Figure 9.7   *Compressing layers of cardboard. The bar on the left is applying a horizontal pressure. The layers are forced together by stiff foam on the top and on the bottom. You can see the elegant zig-zag patterns with a small quasi-periodic ripple as well.*

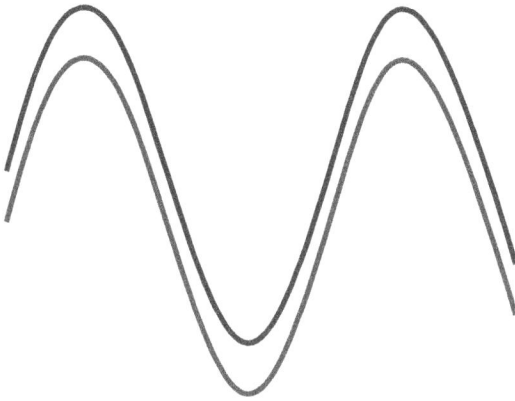

Figure 9.8   *One sine wave does not fit on top of another without creating a void.*

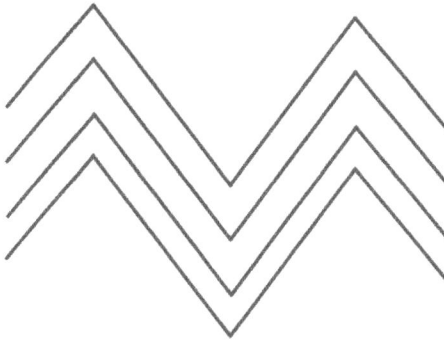

Figure 9.9   *Zig-zag shapes can easily stack together without creating voids.*

So far we have used a differential equation model to predict the folding patterns of rocks and other materials. But there is an elegant *geometrical* explanation of why we get these zig-zag patterns. Rocks typically get folded when they are deep underground and there is a lot of pressure both vertically and horizontally on the layers. The vertical pressures force the rock layers to fit together (just as in the experiments above). So the layers can only deform into shapes which fit together. It is a geometrical fact that sine-waves do not sit one on top of each other (see Fig. 9.8). However, zig-zag shapes do (see Fig. 9.9). Try this yourself with some sheets of cardboard. So if you want to stack a lot of similar shapes one on top of the other then only a zig-zag shape will do.

Figure 9.10 *Chevron folding patterns at Millock Haven, Cornwall.*

So, do rocks deform in the same way was paper? Very much so! Figure 9.10 shows a lovely example of rocks folded into chevron patterns at Millock Haven in Cornwall with a scale provided by a human at the bottom of the cliff. You can clearly see the zig-zag structure, with the layers fitting together perfectly. As an example of a mathematical shape which can be seen in nature these folded rocks are hard to beat. These are very well worth a visit.

If you look at the paper folding examples you will see that there are occasional gaps, or *voids*, in the layers. Typically these are at the sharp corners in the folds. This is because it takes a very large force to bend rock into a sharp corner. If the horizontal force is not large enough then instead of a sharp corner the rock has a slightly more gentle bend, and the rocks don't quite fit on top of each other, leading to a void. Figure 9.11 is an example, again from the rocks in Cornwall.

Using the mathematical model above, we can precisely predict where these voids will occur. Why should we care? Well voids occur in real rock folding, and it is in the voids that minerals are deposited by water running through the rocks. One of these minerals is gold. So mathematics can help in a geological treasure hunt by locating where we might find gold in folded rock layers. Maybe

Figure 9.11 *Voiding patterns in the rocks in Cornwall.*

this might make us all rich! More about rock folding can be found in the classic geological textbook [50].

3. Let there be light: Photonic crystals

Having seen how materials bend, we will now use mathematical models to help us to understand more about their inner workings.

Many materials are made either directly, or indirectly, from rock. Obvious examples we see all around us in construction include lime and gypsum and concrete. The latter material was invented by the Romans and allowed them to build many of their extraordinary pieces of engineering, including docks and aqueducts. Rocks are also closely linked to materials used in communication. It was discovered in the early part of the 20th Century that crystals could be used in radio communication, both to detect signals imposed on radio waves (in the celebrated crystal sets which I used to build) and then to fix the frequency of the same waves. The same technology is used in digital watches to give a highly accurate timepiece. More recently crystals and glass are used in fibre optic technology. It is likely that if you have broadband, that for part of its path, the broadband signal is transmitted along a fibre-optic cable. It is a simple fact that light going down such a cable can not only convey much more information, but can also be transmitted without a lot of loss of signal. This observation led the designers of fibre-optic cables to think about whether it as possible to reduce the attenuation of the signal still further. This would allow very long-distance communication without the need for units to boost the power. This is especially important for the transmission of signals under the ocean. This problem has been solved with the use of some very fancy mathematics in the shape of *photonic crystals*. An example of such a crystal is shown in Fig. 9.12. As you can see the photonic crystal is made up of a fibre-optic cable with a regular lattice of holes drilled into it. To study such crystals we set up the (partial differential) equations for light moving through glass. There are a number of such equations, including the Helmholtz equation and the nonlinear Schrodinger equation for example. The precise equation used depends upon the intensity of the laser light beam, and also the type of crystal used. The Helmholtz equation is the simplest of these, and if u is the intensity of the laser light then it satisfies the partial differential equation

$$u_{xx} + u_{yy} + u_{zz} + \omega^2 u = 0,$$

where ω is the frequency of the wave. When this equation is solved to find u for the above geometry, it is found that at certain frequencies, the holes in the crystal act as miniature resonant cavities. These resonant cavities act effectively to block the light beam going through the middle of the crystal to leak out through the sides. More details of this can be found in [51].

The effect is an effectively loss free cable (provided that it can be manufactured in the first place). Fortunately this is now possible Fig. 9.12 shows an example of just such a photonic crystal. As you can see it is a very complex shape, and until fairly recently it was thought impossible to analyse the behaviour of light in such a crystal without the use of huge, and expensive, computing power. However, advances in manufacturing have gone hand in hand with advances in mathematics. Indeed it is because of the development of new mathematical techniques that we are not only able to analyses these complex shapes, but we

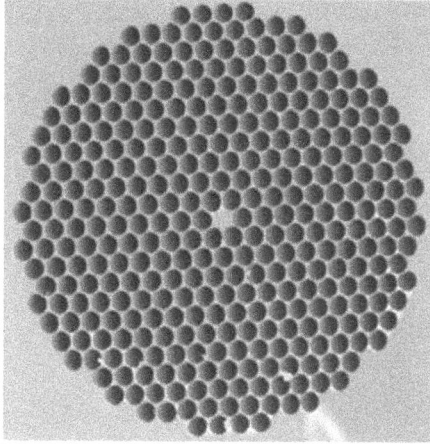

Figure 9.12   *A photonic crystal used to convey information in a fibre-optic cable. (Credit: https:// commons.wikimedia.org/wiki/File:Photonic-crystal-fiber-from-NRL.jpg)*

can also predict their optical performance with some precision. The new mathematical techniques are called multi-scale homogenisation methods and they work by recognising that a material such as a photonic crystal has several scales. It has a micro-scale which is the wavelength of light, an intermediate meso-scale which is the scale of the holes, and a macro-scale which is the width of the whole. (Many other materials have several scales, natural examples are wood, bone and cells.) The material behaves differently at these different scales, and its properties result from the way in which these different behaviours interact. Multi-scale mathematical methods allow us to do this, and thus predict the behaviour of the photonic crystals without the need of a super computer. Multi-scale methods now have many more applications, ranging from weather forecasting to structural optimisation.

4. Crystals and alloys

Crystals

We have so far met photonic crystals. Most engineered materials, including metals and ceramics are in fact polycrystalline. An example of a polycrystal is

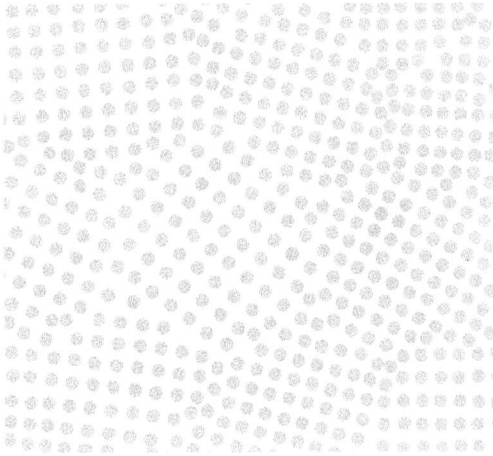

Figure 9.13 *An example of the molecules in a polycrystalline material showing both regular crystalline patterns, and the phase boundaries between the regular regions. (Credit: Edward Pleshakov, https:// commons.wikimedia.org/wiki/File:Crystallite.jpg)*

illustrated in Fig. 9.13. Polycrystaline materials are composed of many small crystalline grains which are separated by boundaries. If, for example, a piece of metal is heated in a furnace, the average size of the metal grains grows. The small grains will disappear, while the big ones will grow, due to changes in the boundaries between individual grains. Exactly the same mathematics can be used to describe this process as describes the evolution of froth on a pint of beer, which we looked at in the last chapter. This raises the important question of whether we can design new materials by predicting the shape of the polycrystals.

Crystals themselves have a strong link to mathematics, with their shape, symmetry, size and strength highly predictable by mathematical methods. A recent example of this is the celebrated 'bucky ball', which is a form of Carbon, in which 60 Carbon atoms are arranged geometrically in the shape of a soccer ball made of 20 hexagons and 12 pentagons, shown in Fig. 9.14.

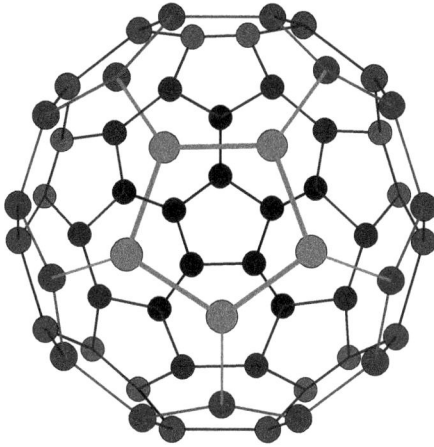

Figure 9.14   *The structure of the Carbon atoms in a bucky-ball. This is the same pattern as in a traditional soccer ball.*

Mathematicians call this shape a *truncated icosahedron.* Crystals take such symmetric shapes, not because they are particularly beautiful (although of course they are) but because they represent minimal energy states for the molecules. This makes them both strong and stable. We saw some lovely examples of crystals in Chapter 1 in the form of snowflakes.

Understanding energy turns out to be the key to looking at many different types of material. The process of doing this mathematically is to find a formula for the energy and to then find the configurations of the material that minimises this energy. This simple idea is the key behind unlocking the secrets of many materials and allows us to predict many varied and exotic shapes of those materials.

Alloys

A material which is mixture of two different metals, or of a metal with another element, is often called a *binary alloy.* Examples of such binary alloys are the steel alloys used in much of manufacturing, the titanium alloys used in aircraft, and brass which has many decorative uses. Bronze (as used in the bronze ages) is an alloy of copper and tin which is harder than either. Typically when making an alloy, the two materials are mixed together when they are hot. The material is then allowed to cool slowly (or annealed). As it cools the separate materials divide into distinct phases and these can give exotic patterns. Understanding these patterns helps us to predict the resulting properties of the material. An example of the patterns found in annealed steel (taken from the Wikipedia article on alloys) is given in Fig. 9.15.

It is possible to describe, understand, predict, classify and control these patterns. One such model for pattern formation in a two-phase material is the celebrated *Cahn–Hilliard equation* which models the concentration c of each phase. This equation is another example of a fourth order partial differential equation which explains the way that the phase changes in time and in space. The Cahn–Hilliard equation takes the (somewhat scary I admit it) following form:

$$\frac{\partial c}{\partial t} = D\nabla^2 \left(c^3 - c - \gamma \nabla^2 c \right).$$

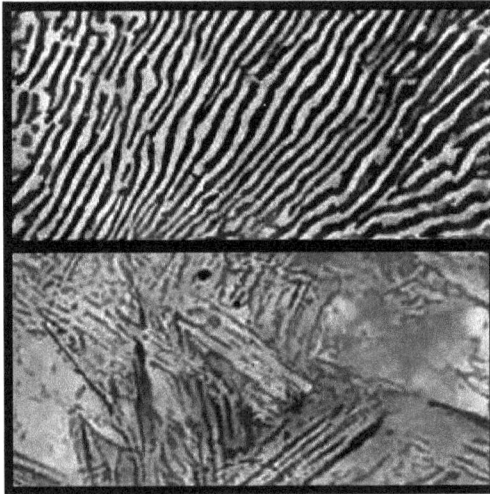

Figure 9.15 *Patterns in annealed steel and quenched steel, from 1911 Britannica plates. (Credit: Osmond and Cartaud, https://commons.wikimedia.org/wiki/File:Photomicrograph_of_annealed_and_quenched_steel,_from_1911_Britannica_plates_11_and_14.jpg)*

This equation has many similarities to the Navier–Stokes equations which we encountered in Chapters 1 and 6. The term $\partial c/\partial t$ expresses the rate of change of the phase with time, and the operators ∇^2 express the rate of change in space. In the solution of the partial differential equation c takes the values of +1 and −1 in each of the two phases, and the expression describes how the phases evolve in time and in space. As you will probably have noticed this equation is nonlinear and cannot be solved analytically. However, it can be readily solved on a computer. A snapshot of the solution is shown in Fig. 9.16. In this figure the red shows where c takes the value of +1 in one phase, purple where it takes the value of −1 in the other phase and the green is the thin transition region where one phase rapidly changes into another. The picture shows the intricate (some would say beautiful) patterns that arise both in the solution of the Cahn–Hilliard equation and the phase separated system that it models.

The Cahn–Hilliard equation has many other applications in polymer science, complex fluids and industrial applications such as the manufacture of steel. It is a major tool in allowing mathematicians to predict the behaviour and form of complex materials which are binary mixtures of two other materials.

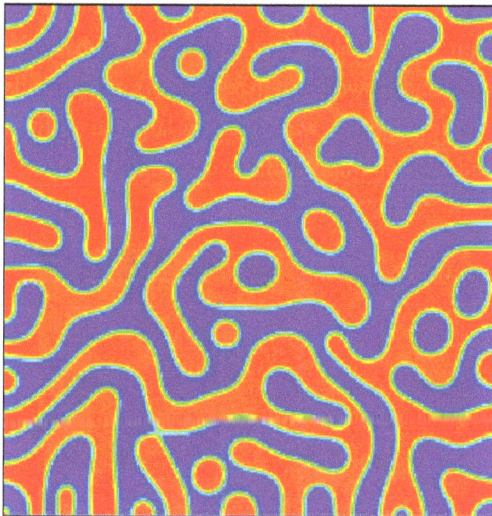

Figure 9.16   *The intricate patterns of the two phases given by the solution of the Cahn–Hilliard equation.*

5. Granular materials and the Brazil Nut Effect

It is time for your breakfast and you reach for your packet of cereal, which is a mix of muesli and nuts. As you pour the cereal into your breakfast bowl you find, much to your surprise, that you get a bowl full of nuts as can be seen in Fig. 9.17. This is odd, because it means that the large nuts have risen to the top of the muesli. This seems to contradict our intuition that larger objects should go to the bottom. What you have seen (literally) is an example of the Brazil Nut Effect. The reason for this is that your cereal packet comes to you having travelled a considerable distance. (To learn more about the movement of food around the world have a look again at Chapter 8.) During its travels it gets shaken up. It is the shaking which is the key to the Brazil Nut Effect. As the cereal is shaken the nuts may move up a little, leaving gaps beneath them. The smaller and faster moving muesli then flows into these gaps. As it does so, so the nuts rise upwards. Eventually they get to the top of the cereal. This is illustrated in Fig. 9.18.

The Brazil Nut Effect is just one of the counter intuitive properties shared by granular materials, which as the name suggests are materials made up of grains. Granular materials are very important in the food industry. For example, they make up the grain in grain silos, they are the basis of baby food, nuts, rice, coffee, chocolate, custard powder and many cereals such as cornflakes. Granular materials also occur naturally, with snow in large quantities being a good example, and sand another. Many of the fascinating patterns seen in sand dunes

Figure 9.17 A breakfast bowl of muesli. Where are the nuts?

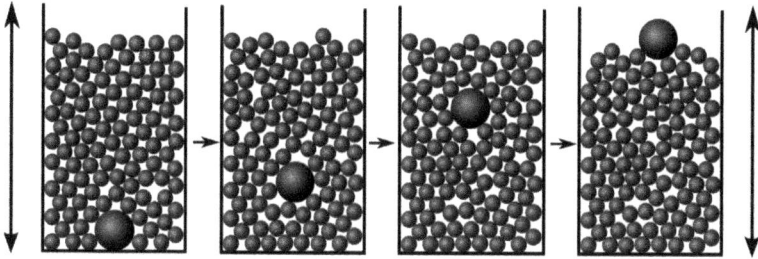

Figure 9.18 *When a granular material is shaken, the large 'nuts' rise to the top in the 'Brazil Nut Effect'.*

are a consequence of it being a granular material [52]. Another application of granular materials is in the manufacture and storage of paint powders, fertilizer, cement and even sand. Mathematics is used to predict how granular materials can flow. In some sense, granular materials do not constitute a single phase of matter, but have characteristics similar to solids, liquids or gases, depending on the average energy per grain. However, in each of these states granular materials also exhibit properties which are unique. Indeed, granular materials also exhibit a wide range of pattern forming behaviours when excited (for example, vibrated as above, or allowed to flow). Different behaviours arise depending upon whether the grains can move freely over each other, or if they can get jammed. The latter is certainly what you don't want in a grain silo. As such, granular materials under excitation can be thought of as an example of a complex system, and we will meet these again in Section 7. Mathematics plays a critical role in predicting whether jamming will or will not occur, and by doing this it helps to feed the world.

6. Harry Potter's cloak of invisibility

Many of us will have watched *Harry Potter* films, or maybe are fans of *Star Trek*. Both of these contain devices for making you invisible. In the case of *Harry Potter* a special cloak, and for *Star Trek* the infamous Cloaking Device [53]. Once thought just to be the product of the imagination, the modern science of meta-materials is coming closer to the invention of a true invisibility cloak, that can render the wearer invisible when they are watched from several different directions. Early invisibility devices used a combination of lenses and mirrors to bend the light around an object to make it appear invisible. An example of such an arrangement to make a potato disappear is shown in Fig. 9.19. Devices like this were used to produce special effects in Victorian Music Halls. Making these devices required a knowledge of geometrical optics, and in particular the mathematics of angles and trigonometry. They make use of the fact that to a good approximation light travels in straight lines. Therefore, the paths that it takes can be described by the sort of geometry that would have been very familiar to the Ancient Greeks.

An example of such a cloaking device was invented in the University of Rochester and is called the *Rochester Cloak*. This uses four lenses to direct light rays around the objects to be cloaked. However, these methods of cloaking using lenses and mirrors could only work from one direction. There has been intensive research into finding materials which can cloak someone from many directions. One example are stealth aircraft which have a special shape, and are covered in special materials, to make them nearly invisible to Radar. Researchers in the US have now invented a digital cloaking device that works from many different directions, and despite some shortcomings suggests such technology may be closer to practical uses in the real world than you might think. This system calculates the direction and position of the light rays so they can be properly displayed as if they were unobstructed. As a result, the area behind

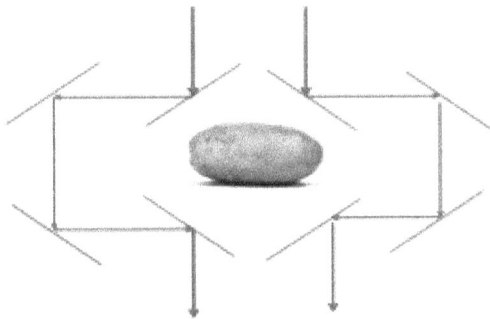

Figure 9.19 *An arrangement of mirrors which would render the potato in the middle invisible.*

the display is effectively cloaked. As the viewpoint shifts, the image on the display changes accordingly, keeping it aligned with the background.

Invisibility cloaks themselves rely on meta-materials, which as we described in the introduction, are a class of material engineered to produce properties that don't occur naturally. Light is electromagnetic radiation, made up of vibrations of electric and magnetic fields. Natural materials usually only affect the electric component, but meta-materials can affect the magnetic component too, expanding the range of interactions that are possible. The meta-materials used in attempts to make invisibility cloaks are made up of a lattice with the spacing between elements less than the wavelength of the light, which can then be bent by the material.

A related development has been the design of clothing which is not only flexible to fit the body, but also has the properties of the LCD screens described above. By using these materials we can envisage a future where the colour of a dress or a suit can be instantly changed to meet the needs of an event, and can even display moving writing if a sponsor desires it. Maths meets fashion indeed.

7. Getting electrical in a complex way

Suppose that you want to construct a building at an airport. If that building reflects radar waves then you have a problem. The radar at the airport, or in the aircraft landing or taking off there, will reflect off the building at all sorts of angles. The result is that the radar system gets confused and the radar displays get cluttered. Therefore, it is a good idea to cover buildings with a material that absorbs the radar waves. The same principle applies if you want to make a stealth aircraft invisible to radar (rather like the *Harry Potter* cloak). During World War II it turned out that nature had already solved this problem The famous de Havilland Mosquito was made of wood, and this made it almost invisible to the German radar systems. Now, the challenge is to make a synthetic material with the same radar absorbing properties. To do this we make use of the new ideas of *complexity* and of *emergent properties*. Loosely speaking, a complex material is something which is made up of many different materials. An example of such is a composite material such as carbon fibre, which is used in many different applications, from aircraft engines to fishing rods. The key feature of such a complex material is that its properties are much more than the sum of the different component properties. Instead they reflect the way that these different properties *interact*. Or more poetically, the whole is greater than the sum of the parts. Such properties are then said to *emerge* from the interactions, and may be very different from the original properties of the component materials. The only way to find out what these emergent properties are is is to use quite sophisticated mathematics.

The question of designing such a radar absorbing material came to our team at Bath, and it was decided to make a complex meta-material combining the different electrical properties of Aluminium Oxide and Titanium Oxide. On the technical side, the first of these behaves like a resistor, and the second like a capacitor. Resistors have electrical properties, which are independent of the frequency of the electrical waves, which pass through them. In contrast, capacitors conduct electricity well at high frequencies, and badly at low frequencies. To create a mathematical model of the complex material we took a random mixture of these two different materials mixed in careful proportion. Figure 9.20 is a picture of the resulting mixture. The patterns

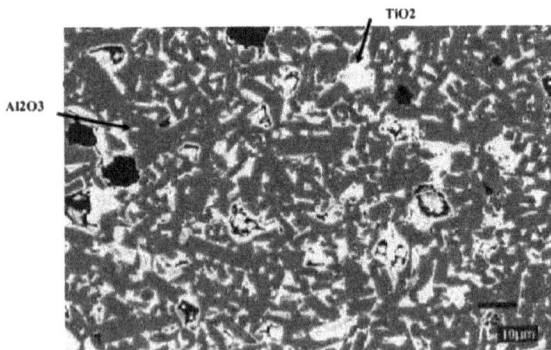

Figure 9.20   *A composite material which is made up of Aluminium Oxide (grey) and Titanium Oxide (white). The patterns made by these materials are similar to those shown in Fig. 9.16. These two materials have very different electrical properties which combine to give an emergent electrical behaviour of the composite material.*

of the two materials are very similar to those predicted earlier by using the Cahn–Hilliard equation.

As you can see this material is very complex, and the electrical properties appear very hard to analyse. So we had to think carefully about how we would construct a mathematical model of this situation which could be used to make predictions. We realised that what was key to this situation was that the Titanium Oxide behaved like a lot of small resistors and the Aluminium Oxide like a lot of small capacitors. We could model the way that these connected together by using network of resistors and capacitors, as in Fig. 9.21.

This network is rather like the electrical supply networks that we met in Chapter 7. The main differences being that in a real material there are many more connections than in a supply network, and whilst in a power network we know exactly what joins each point, in the material this is much less certain.

Now we use a trick often employed in using a mathematical model to study complex problems. In this approach, instead of trying to exactly reproduce the parts of the material, we instead think about using a *statistical approach* to model the complexity. To do this we don't try to predict in our model exactly which of the connections in the network is a resistor or a capacitor.

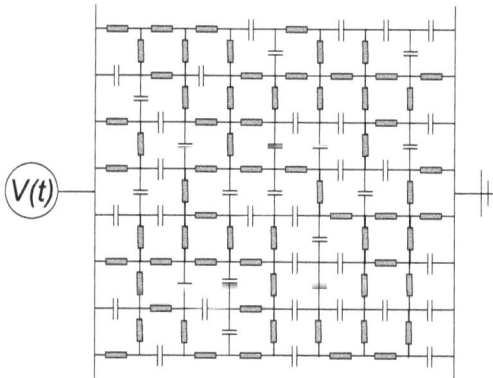

Figure 9.21   *A random network of resistors (shown as solid blocks) and capacitors (shown as parallel plates) which models the electrical behaviour of the composite material when subjected to a voltage V(t).*

Rather than this we say that the connection is a capacitor C *with probability* p and a resistor with probability $(1-p)$. So to decide whether a link is a capacitor or a resistor we toss a (biased) coin, and depending on which way it falls we make our choice. If we repeat this for every link in the network we get what is called *a realisation* of the material. If we do this a second time we will get a different realisation with a different arrangement of capacitors and resistors. If we do this a lot of times (which is not difficult on a computer) we can then find the *statistics* of the electrical properties of the resulting network, in particular the statistical behaviour of its conductivity, which is a measure of how well it can conduct electricity.

An interesting feature of the conductivity σ of the network is that it depends on the frequency w of the alternating voltage $V(t)$. The conductance of a resistor is $1/R$ and of a capacitor is $i\,\omega C$.

The conductance of a resistor does not depend upon frequency, but the conductance of a capacitor gets larger with increasing frequency. The use of the imaginary number i (which we met in Chapter 5) means that when the capacitor conducts electricity it also changes its phase.

The network of resistors and capacitors itself can be represented by a matrix with random coefficients, and its statistical properties found by the theory of such random matrices. This turns out to be a very powerful technique for predicting the properties of the original material and gives excellent results. (We use the same idea of representing complex processes by random variables in many other applications, including studies of the behaviour of the stock market, the movement of crowds of people and in the climate problems that we looked at in Chapter 6.)

One way to find the conductivity is to use the process of *logarithmic mixing*. If you suppose that the material passes a current of frequency ω then from above we know that the conductivity of the resistors in the network does not depend upon the frequency and is a constant $1/R$, and the absolute value of the conductivity of the capacitors increases with frequency and is given by ωC.

The random matrix theory then predicts that if there is a good mixture of resistors and capacitors, then the logarithm of the overall conductivity σ of the material, is a mixture of the logarithm of the conductivity of the capacitors and the resistors in direct proportion to the amounts of each in the material. Or to put it in mathematical terms

$$\log(\sigma) = p\log(\omega C) + (1-p)\log(R).$$

This tells us that the conductivity σ changes with the frequency ω according to the formula:

$$\sigma = C^p R^{1-p} \omega^p.$$

This can be tested, and it works. Indeed the predictions from this model agree very well with experiment (see Step 6 of the modelling process), and are now used (see Steps 8 and 9) to help design the materials that the airports need for its future buildings.

Random matrix methods are now finding many other applications in the design of other complex materials [54], as well as in communication theory, and even in the celebrated Riemann Hypothesis.

8. Paper and origami

One of my favourite materials is the humble sheet of paper. Perhaps because it is so commonplace and ordinary that we forget that the invention of paper by the Chinese has possibly had the biggest effect on human civilisation of all materials. The invention of paper allowed both the rapid transmission of ideas, and the ability to store knowledge for future generations. It has been around for a lot longer than modern electronic storage media, and has the advantage that words written on paper can be read many years later, whereas there is no guarantee that a particular electronic means of storing data will still be usable fifty, or even ten, years into the future. Who still uses floppy discs for example? Paper is typically made from wood pulp which is added to boiling water and then laid out in large sheets and left to dry, before being put onto giant rollers as we can see in Fig. 9.22.

Figure 9.22 *Making paper in a factory. (Credit: https://www.shutterstock.com/image-photo/paper-production-plant-on-towels-fragment-510317089)*

How big is a sheet of paper?

Mathematics is certainly important in regulating the heat and drying process for making paper, but it really comes into its own when it comes to designing the final shape of the paper that we use. In recent years most countries (apart from those in North America) have adopted the A0, A1, A2, A3, A4, A5, ... standards for the shape of the paper that they use. These **all** have the same, carefully chosen, proportions illustrated above. If you take a sheet of A0 paper and fold it in half along its longest side, then you get A1, fold this in half and you get A2, then A3, then A4 etc. The sheet of A0 paper has area of exactly 1 m^2, so the area of A1 is 1/2 m^2 and A4 is 1/16 m^2. However, every such sheet of paper has *exactly the same shape*.

One very nice feature of this sequence of paper sizes is the way that they fit together. If you take the sheets A1, A2, A3, A4, A5, A6, A7, A8, etc. and find their combined areas then you get the series

$$1/2 + 1/4 + 1/8 + 1/16 + 1/32 + 1/64 + 1/128 + 1/256 + \cdots$$

This is an example of a geometrical series in which the ratio of the successive terms is 1/2. If we set

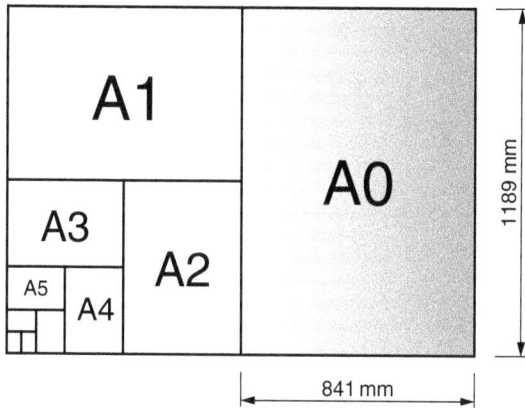

Figure 9.23   *How you can make A0 paper by combining A1, A2, A3, ...*

$$S = 1/2 + 1/4 + 1/8 + 1/16 + 1/32 + 1/64 + 1/128 + 1/256 + \cdots$$

Then, multiplying by 2 we get

$$2S = 1 + 1/2 + 1/4 + 1/8 + 1/16 + 1/32 + 1/64 + 1/128 + 1/256 + \cdots$$

It is clear from looking at these two infinite series that

$$2S = S + 1.$$

Hence, we get

$$\boxed{S = 1.}$$

What this means is that if you combine all possible sheets of paper A1, A2, A3, ... together, then you get a sheet of paper of size 1 m². This is the same area as a sheet of A0 paper. We can see this in Fig. 9.23.

Only one shape of paper has the property, and this can be worked out mathematically. Suppose the proportions of a piece of A paper are $x/1$ with x being the length of the longest side, and without loss of generality we take the length of the shortest side to be 1. If you fold this paper in half this divides it into two equally sized rectangles. On these smaller rectangles the length of the longest side is now 1 and the length of the shortest side is $x/2$. The proportions

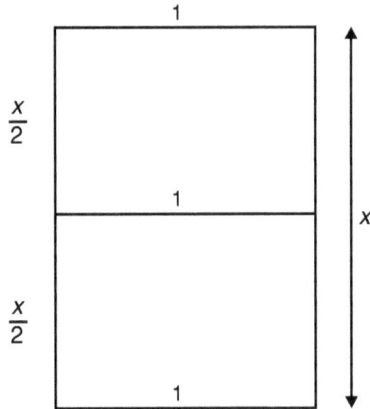

Figure 9.24   *The proportions of a rectangular sheet of A-sized paper. These must be the same as the proportions of the rectangle that you get when you fold this in half.*

of the smaller rectangle are then given by the ratio of the longest side to the shortest side, which is now $1/(x/2)$. This is illustrated in Fig. 9.24.

In A-sized paper these two ratios must be the same. Thus, x must satisfy the equation:

$$\frac{x}{1} = \frac{1}{x/2} = \frac{2}{x}.$$

Rearranging this equation gives $x^2 = 2$. Or in other words

$$x = \sqrt{2} = 1.4142135623730950488\ldots.$$

So, the length of the longest side of sheet of A paper is exactly the square root of 2 times the length of the shortest side. Now we will test our model to see if we can work out the size of a sheet of A4 paper.

We know that the area of A0 paper is 1 metre squared. Therefore, the area of A4 paper must be 1/16 of a metre squared. Suppose that the shortest side of A4 paper has length L. Then the longest side has length $\sqrt{2}\, L$ and the area of a sheet of A4 is. $\sqrt{2}\, L^2$.

Therefore, L satisfies the equation:

$$\sqrt{2}\, L^2 = 1/16.$$

Solving this equation gives

$$L = 0.2102 \text{ m} = 21.02 \text{ cm}.$$

If you take an actual sheet of A4 paper, then its (officially designated) longest side is 29.7 cm and its shortest side is $L = 21$ cm, so its proportions are $1.41428{:}1$ and L is very close to the value above. If you don't believe me then try measuring these dimensions for yourself.

The square root of 2

The famous irrational number $\sqrt{2}$ is the length of the diagonal of a unit square, and it is important in calculating lengths and areas. In this context it was known to the Babylonians, and they computed a very close approximation to it given by

$$305470/216000 \approx 1.414213.$$

This approximation was recorded on the YBC 7289 clay tablet dated to around 1700 BC. It was obtained by an iterative procedure, still used today, in which if x_n is an approximation to $\sqrt{2}$ then you can get a better approximation by calculating

$$x_{n+1} = \frac{1}{2}\left(x_n + \frac{2}{x_n}\right).$$

If you start with $x_1 = 1$ then this iteration very rapidly converges to the square root of 2.

The number $\sqrt{2}$ has the beautiful expression as a continued fraction given by:

$$\sqrt{2} = 1 + \cfrac{1}{2 + \cfrac{1}{2 + \cfrac{1}{2 + \ldots}}}$$

and this continued fraction expression can also be used to calculate it to a high degree of accuracy. However, any such calculation can never be perfect, as the ancient Greeks showed that $\sqrt{2}$ was irrational and could not be expressed as the ratio of two whole numbers. This caused consternation at the time it was discovered and changed their whole approach to mathematics, which up to that time had assumed that all numbers were simple ratios. The proof of the irrationality of $\sqrt{2}$ can be found in many text books and on the Internet, and I won't repeat it here. The first proof was probably the one published in Euclid's text book. This proof also appeared in the 1978 Royal Institution Christmas Lectures by Professor Sir Christopher Zeeman. His demonstration is possibly the first time that a mathematical proof had been shown on mainstream TV.

The square root of 2 is one of the most important numbers in the whole of mathematics. It appears everywhere from trigonometry and geometry to calculus and differential equations. Its many applications include estimating the mean of the electrical power we looked at in Chapter 7, and in the frequency ratio of a tritone interval in music. It is great to know that whenever you use A4 paper you are looking at this wonderful number.

Origami

At the start of this chapter we looked at the mathematics of folding cardboard, and then rock. Once you have a sheet of paper it is also natural to want to fold it. The Japanese have been doing this for centuries and have developed the wonderful art of origami creating a huge variety of designs of animals, flowers, people and decorations, all folded from a single square sheet of paper, without any cutting. We have already seen above that there is a close link between folding and mathematics. Origami can also be used to solve mathematical puzzles. Two of these are the questions of trisecting an angle, and doubling the volume of a cube. These were posed by the Greeks and we now know that they cannot be solved by using a ruler and a compass on a piece of paper. However, once you are also allowed to *fold* the paper then they become soluble. A lovely account of how to use origami to trisect an angle is given in the *Plus Magazine* article [56].

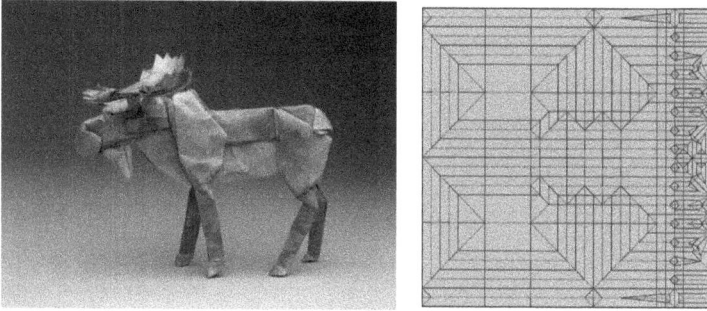

Figure 9.25 *The bull moose fold, designed by the origami mathematician Robert Lang [55]. The fold is on the left, and the folding pattern is on the right. (Credit: With thanks to Robert Lang)*

In recent years, mathematicians have become increasingly interested in origami. They ask the questions of what shapes can in theory be folded from a piece of paper, and also how these shapes can be created. Out of this study have come some incredible new origami designs, which are a true fusion of mathematics and art. One of the leading exponents of these new designs is the wonderful mathematician/origamist Robert Lang (others are Erik Demaine and Tom Hull), and both the theory and examples of his work can be found in [55]. Figure 9.25 shows one of his designs, the Bull Moose. On the left is the origami creation and on the right is the set of folds on the square sheet of paper which lead to the moose design. Robert Lang also uses his origami skills to design the solar panels and radio antennae for satellites, which have to be unfolded once they get into space and we will see examples of these in Chapter 10.

I challenge anyone, who sees these folding patterns not to say that mathematics is a true bridge between the arts and the sciences!

10

Mathematics goes into space

1. Introduction

As we saw in Chapter 1, one of the earliest, and perhaps one of the most important, of the applications of the ideas of mathematical modelling, came from studying the Solar System. Mathematical modelling continues to be a powerful tool in opening up the possibilities of outer space and in showing us the possibilities for the future, with an example of the use of a mathematical model applied to a satellite shown in Fig. 10.1. It is thus fitting to end this book by looking at the applications of mathematics to studying space.

Outer space has fascinated human beings ever since we developed enough intelligence to ask questions about the world that we lived in. Long thought to be the realm of the Gods, space was considered to be beyond our comprehension, and studying it was even blasphemous. However, brave pioneers looked up and started to study carefully what they saw. The resulting understanding of the nature of space has had profound impacts on human civilisation. As agriculture developed and crops had to be sown at the right time of the year, so it became profoundly important to understand the seasons and the motion of the Sun. Knowing the motion of the Sun also led to an

Figure 10.1   *A satellite orbiting the Earth. Without mathematical models used to predict its orbit, it would never have got into space. (Credit: https://www.shutterstock.com/image-illustration/satellite-communications-earth-reflecting-solar-panels-150180452)*

understanding, and measurement, of time. Later it was realised that there were other heavenly bodies such as the stars, Moon and planets and their motion was studied and analysed in terms of mathematics. Understanding these was very important in the development of navigation (from the earliest seafarers onwards) and (perhaps less usefully) in the growth of astrology.

Research into space has been an enormous stimulus to the growth of science and technology in general, and of mathematics in particular. Perhaps the most important example of this is the development of calculus. Space research has also greatly benefited from developments in mathematics. Without mathematics we would never have had the Moon landings for example, or to have been able to launch any satellite. But the application of mathematical models linked to space was also the foundation of our ability to predict the seasons, eclipses, to navigate on the oceans, and to communicate reliably over vast distances. It is certain that mathematical models will play a dominant role in the future of all and any technology that goes into space.

Space is now a very big business with a total value of commercial investment on space technology estimated at $50 billion! This fact was recently recognised by the UK Government that identified Space Technology as the second of their eight great technologies. To give some idea of the scale, in 2016 there were 85 rocket launches into space (of which 79 were successful), there were 87 in 2015 and 92 in 2014. Most of these launched satellites into near Earth orbit. It is currently estimated that there are 4,256 such satellites in orbit of which 1,419 are still active (although there may be secret satellites that we don't know much about. See the editorial in *Maths Today* [57] for more details of the scale of this activity). We then can add to this the rarer, but very glamorous, activity of sending satellites (and even people) to other bodies such as the Moon, Mars, comets, the distant planets and beyond. I grew up in the 1960s and can testify to the huge impact that the Moon landings had on my generation, including awakening and stimulating my own interest in science and maths!

It is the day to day work of satellites which is making a huge difference to the modern world. This includes the transmission of huge amounts of data, GPS navigation, remote sensing, weather observations, relaying mobile phone messages, agricultural monitoring, whale spotting (yes, it can be done) as well as giving us a window to space from space. It is hard to imagine how we would function without all of this technology in space, most of which would not work at all if it was not for the application of a wide variety of mathematical ideas.

In this chapter we will start by looking at the early days of space technology, before we sent anything into space itself. We will then look at the technology behind orbital satellites and how satellites (and also Apollo) were guided beyond the orbit of the Earth to the moon and the planets. Next we will take a look at deep space showing how Einstein's General Relativity not only describes the structure of the Universe, but is also important in the functioning of GPS satellites. Finally we will come back closer to Earth with a bit of space origami.

2. Space from Earth

As I said in the introduction, space has been studied since the earliest times, but it was the Chinese and the Babylonians who made the first serious studies of it by using mathematical models. A reason for this was simple; it was clear that the heavens governed the seasons, and it was the understanding of the seasons that was vital for agriculture. Indeed in the Book of Genesis we read

> *And God said, Let there be lights in the firmament of the heaven to divide the day from the night; and let them be for signs, and for seasons, and for days, and years.*

We now know that the changing seasons are due to the tilt of the Earth's axis, and the change in the angle that the Sun makes to the horizon during the year. Whilst this was not known to the ancients, it was clear to them that the Sun governed their lives. In order to work out when was the best time to plant crops they had to predict when the Sun would be in the right part of the sky, and to do this they needed an accurate *calendar*. Creating this was an early application of the mathematics of space to human life. To produce a calendar it was important to know the length of the year. This is not as simple as it looks as the year is 365.2422 days long, which is not an easy number to calculate. The Babylonian calendar used the value of 365 days, which quickly lead to a loss of accuracy. The Egyptian and later Julian calendars used instead the value of 365.25 days. To achieve this the leap day was introduced so that there are 365 days in a normal year, and in a leap year (every four years) there are 366 days. Close but not quite close enough.

The Gregorian calendar improved on this by omitting leap years that fall on 100 years. Because 97 out of 400 years are leap years, the mean length of the Gregorian calendar year is 365.2425 days, which is much closer to the value of 365.2422 and is the one in main use throughout the world. As a result the seasons can be predicted with great accuracy. Further mathematics was needed in order to account for the motion of the Moon and (amongst other things) to calculate the date of Easter Sunday, which was defined in 325 AD by the Council of Nicaea to the first Sunday after the first full Moon occurring on or after the March equinox. It is said that this tricky calculation (it is difficult because the period of the orbit of the moon is a complex fraction of the year) kept mathematics alive during the middle ages. Similar (but harder) calculations are needed to determine the tides, and the need to do these were one of the factors leading to the invention of the analogue computer.

Mathematics was also employed by the ancients to calculate the dates of solar eclipses. The Chinese realised that these occurred with a certain regularity and could be predicted by exploiting patterns in number sequences related to the regular periods of the Sun and of the Moon. The *Chinese Remainder Theorem* in number theory, the earliest known statement of the theorem, appeared in the 3rd Century book *Sunzi Suanjing* by the Chinese mathematician Sunzi, and asked the question.

What numbers have remainder 2 when divided by 3, remainder 3 when divided by 5, and remainder 2 when divided by 7?

I will leave it to you to find the answer. The Chinese remainder theorem led to the mathematical theory of congruences. Much later, in the 1801 in his

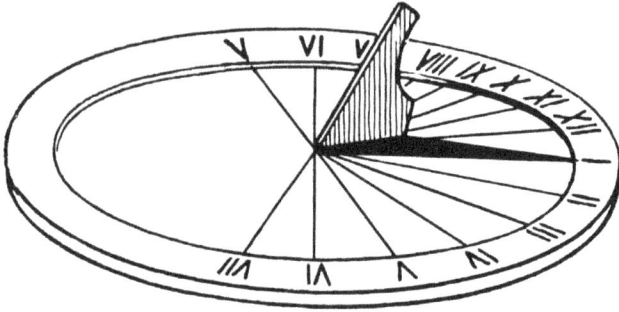

Figure 10.2 *A classical sundial showing the angled gnomon and the lines for each of the hours. (Credit: Pearson Scott Foresman, https://commons.wikimedia.org/wiki/File:Sundial_(PSF).png)*

book *Disquisitiones Arithmeticae* [58], arguably the greatest ever mathematician, used the Chinese remainder theorem on a problem involving calendars, namely, "to find the years that have a certain period number with respect to the solar and lunar cycle and the Roman indiction". All of this shows that pondering the questions of space can not only be helped with mathematics but also leads to great mathematical discoveries.

The observation of space from Earth has had a number of other consequences of direct benefit to humankind. One of these has been in telling the time. It is undeniable (though slightly sad) knowing the time at any point in the day is of huge importance to the running of a civilised society. The time of the day can be determined by looking at the path of the Sun through the sky so that (in the Northern Hemisphere) it rises in the East and sets in the West and is most Southerly, and highest in the sky, at noon. It was realised early on that the Sun's movements were very regular, and understanding this led to the invention of the sundial which tells the time by casting a shadow. The classic sundial design is illustrated in Fig. 10.2.

A lot of mathematics has to go into the design of the classic sundial, including the angle of the pointer (or gnomon) which is the same as the latitude of the user, and also of the angle of the various lines for the times. More advanced sundials, such as the one illustrated in Fig. 10.3 (and which can be found at Simon Fraser University in Canada) have a more complicated design of gnomon

Figure 10.3 *The sundial at Simon Fraser University, Canada. The pointer (gnomon) has a special shape which corrects for the equation of time, which is the variation in the length of the day throughout the year from the mean value of 24 hours.*

(in the shape of an Analemma) which accounts for the fact that the length of the day (measured from noon to noon) is not exactly 24 hours but varies by up to 15 minutes from the mean value of 24 hours throughout the year. (This is why GMT is Greenwich *Mean* Time.) My own favourite design of sundial is the Analemmatic Sundial (see how to build one in [59]) which has the shape of an ellipse (see later) and uses a (vertical) human being to cast the shadow. This is shown in Fig. 10.4 which shows the Jubilee analemmatic sundial outside the Houses of Parliament in the UK. Analemmatic sundials can also be found in many school playgrounds (modesty forbids me to say who designed these).

Sundials in general were used heavily to tell the time until the invention of the mechanical clock, and used correctly they can be very accurate indeed. We get the concept of *clockwise* from the fact that in the Northern Hemisphere the shadow on a sundial goes clockwise around the dial.

3. Space close to Earth

Placing a satellite in orbit

Now, in the 21st Century, instead of just looking at space from the Earth, we are also able to look at the Earth from space. The reason that we can do this

Figure 10.4 *An analemmatic sundial with a human figure casting the shadow, outside the Houses of Parliament in the UK.*

is, of course, that we can send satellites into space with cameras on them. Such satellites are both sent into space, and their location changed in space, through the use of rockets.

In the *Principia* [67], Newton considered firing cannon balls into space, reasoning that if a cannon ball was fired from a high enough mountain, at a high enough speed, then it would orbit the Earth (see Fig. 10.5). Strictly speaking Newton was completely correct in his reasoning (which shows just how far ahead of his time he was), but for 'Newton's Cannonball' to work, the cannon ball not only has to go very fast but it has to get above the Earth's atmosphere to avoid problems with air resistance. To do this, you need a rocket.

The idea of using a rocket to launch a satellite into space was (arguably) first proposed as a serious scientific endeavour by Oberth in the 1930s [57] and

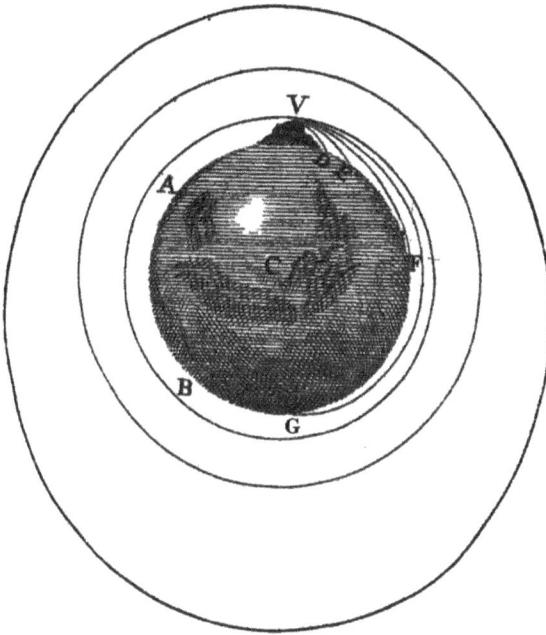

Figure 10.5 *Newton's Cannonball as illustrated in the Principia [67]. If a cannonball is fired from a tall mountain, then for slow speeds it will fall back to the Earth. But for faster speeds it will be able to orbit the Earth. Later in this chapter we will work out just how fast it needs to go to do this.*

then developed by many others since, including Werner Von Braun working first in Germany, and then America, and also Sergei Korolev for the USSR. As is well known the first satellite into space was Sputnik, launched in 1957 (and causing a great shock to the USA in the process). Much the same technology is still in use today. The launch of a satellite on a rocket consists of a short period of powered flight during which the satellite is lifted above the Earth's atmosphere and accelerated to orbital velocity by the rocket, assisted by the 0.5 km/s rotational velocity of the Earth. Usually such a rocket has multiple stages, with large fuel bearing stages discarded early on during the flight. The powered part of the flight finishes when the rocket's last stage burns out. At this point the satellite begins its free flight subjected (at least initially) only to the gravitational pull of the Earth.

When the rocket is launched it initially takes off vertically, and then bends in the direction of the Earth's rotation to insert the satellite into a horizontal orbit.

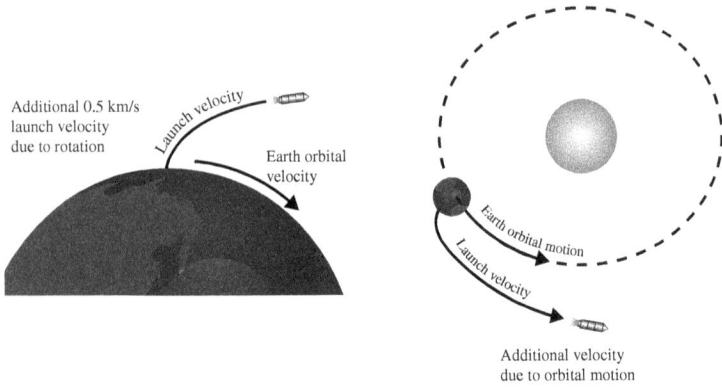

Figure 10.6 *Launching a rocket into space, with some help from the Earth's rotation.*

Such an orbit is shown in Fig. 10.6. The essential physics of this process was first identified by Galileo. When a body is in a circular motion of radius r and it has velocity V then it has a constant *centripetal acceleration* a *towards the centre of the circle* which is given by

$$a = \frac{V^2}{r}.$$

Provided that the satellite is close to the Earth it will have an acceleration of $g = 9.8$ ms^{-2} towards the centre of the Earth due to the action of the gravitational attraction of the Earth.

Thus a near-Earth circular orbit can be achieved provided that $a = g$ and $r = R$ so that

$$V = \sqrt{gR}.$$

The radius of the Earth is $R = 6,371$ km, and provided that the satellite is in a near-Earth orbit, we can take r to equal this value. It then follows that $V = 7.9$ km per second. This is called the *insert velocity* and is a very high speed.

For Newton's Cannonball to work he would have had to have a very powerful cannon indeed. Even a strong cannon would have trouble in firing a cannonball at a speed of more than 100 m per second which is far too slow. This speed could be obtained by using the large rockets developed after World War II, and so satellites finally became possible.

More generally, the gravitational acceleration g due to the Earth at an orbital radius of r is given by

$$g = \frac{GM}{r^2},$$

where G is the Gravitational constant $G = 6.67 \times 10^{-11}$ m³ kg⁻¹ s⁻², and $M = 5.972 \times 10^{24}$ kg is the mass of the Earth. The insert velocity for a circular orbit with a radius of r is then given by

$$v = \sqrt{\frac{GM}{r}}.$$

If the insert velocity differs from the value v given above then the satellite will typically take an elliptical orbit rather than a circular one.

Geostationary satellites

The further away a satellite is, the slower it needs to travel to stay in orbit. Now the Earth takes 24 hours or 86,400 seconds to rotate. The angular velocity of the rotation of the Earth is then given by $\omega = 2\pi/86{,}400 = 7.27 \times 10^{-5}$ radians s⁻¹.

If the satellite travels at the right distance r then its angular velocity will exactly match that of the Earth's rotation. We find this by solving the equation

$$\omega = \frac{v}{r} = \sqrt{\frac{GM}{r^3}} = 7.27 \times 10^{-5} \text{ rad s}^{-1}.$$

This gives $r = 42{,}000$ km. A satellite at this radius from the centre of the earth is then in a *geostationary orbit* and will appear stationary if viewed from the surface of the Earth. This is very useful for a satellite used to relay communications around the world, such as the Telstar satellites launched in the early 1960s. The science fiction author Arthur C. Clarke (of 2001 fame) predicted the use of geostationary orbit for communications satellites in the prophetic paper [60] published in *Wireless World*. In his honour a geostationary orbit is also known as a *Clarke Orbit* and the many satellites in a geostationary position orbit in the *Clarke Belt*. Once in a geosynchronous orbit a satellite can be used to communicate rapidly with the Earth both to relay signals starting from Earth (such as TV programmes or mobile phone messages) from one side of the planet to another (which could not have been achieved before satellites due to the curvature of the Earth), or to transmit data gathered by the satellite itself, such as weather information, remote sensing of the land and sea, GPS signals, or images from deep space. This huge amount of data is having a huge impact on technology and indeed on the whole of society!

Changes to an orbit

It is unusual for a satellite, or indeed any space vessel, to stay in the same orbit throughout its working life and it has to be transferred from one orbit to another, possibly to a planet distant from the Earth. For example, we may need to transfer from an initial parking orbit to the final mission orbit. To change the orbit of a space vehicle, its velocity must be changed through a series of rocket burns which act as impulses to change the momentum of the satellite in its orbit. Such operations require careful mathematical planning to be successful. In particular satellites need to be guided, navigated and controlled in order to move on a prescribed trajectory. This is typically done by a series of carefully calculated rocket burns, designed in advance by computer optimisation methods.

During these manoeuvres great care must be taken to avoid using the available fuel for the mission. The amount of fuel used is given by the *classical rocket equation*. In this we have

$$m_{fuel} = m_{initial}\left(1 - e^{-\frac{\Delta V}{Isp}}\right).$$

Here, m_{fuel} is the amount of fuel burnt, $m_{initial}$ is initial mass of the spacecraft, I_{sp} is the specific impulse of the (rocket) propulsion system (due to the expulsion of a propellant at speeds in the realm of 4,000 m/s) and ΔV is the change in the velocity. It is the velocity change which represents the cost of the manoeuvre. If this is too large then too much of the fuel will be burned. Thus the task of a mission designer is to make the change in velocity due to the rocket burns as small as possible. We will now see how this can be achieved at minimal cost.

4. The Solar System and beyond

Kepler and Newton's laws

As I said in the introduction to this chapter, one of the more glamorous aspects of space technology (and certainly the one that we most think about when we think of space) are the missions of Apollo to the Moon and of satellites (such as Voyager 1 and 2) to Jupiter, Saturn and beyond. These journeys cover vast distances, and are accomplished with the small amounts of fuel carried on the spacecraft. The only reason that this is possible is due to an understanding and application of a mathematical model of the Solar System based upon Newton's law of gravitation, together with many careful calculations. The story behind these calculations involves the whole quest to understand the motion of the planets in the Solar System, and starts with the Greek mathematician Apollonius of Perga in the 2nd Century BC. Up until his work, the curves studied by the Greeks were either straight lines or they were circles. Whilst they were able to do a lot of geometry with these curves, it greatly restricted the sort of problems that they could study. Apollonius' great breakthrough was to introduce a whole new class of curves obtained by taking sections through a cone, leading to the term *conic sections*. The resulting curves are illustrated in Fig. 10.7 and comprise the closed curves of the circle and the ellipse, and the open curves of the parabola and hyperbola.

These curves were studied in detail, and the equations for them derived, so that in Cartesian coordinates, the equations for the ellipse and hyperbola are respectively

$$\frac{x^2}{a^2}+\frac{y^2}{b^2}=1, \quad \text{and} \quad \frac{y^2}{b^2}-\frac{x^2}{a^2}=1.$$

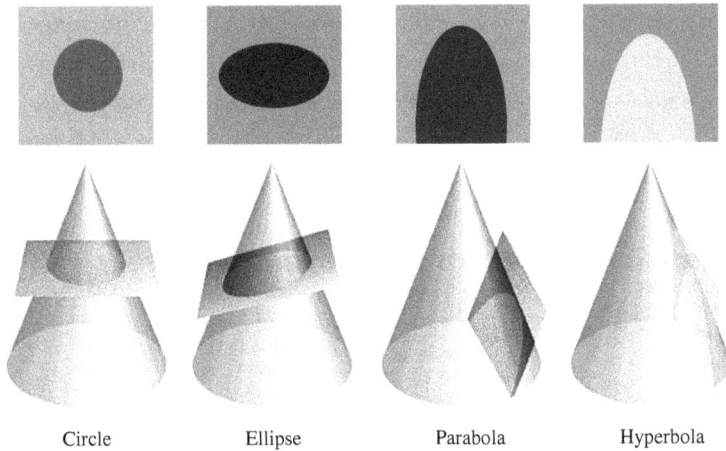

Figure 10.7 *The basic conic sections as discovered by Apollonius.*

Or, in polar coordinates (r, θ) we have

$$r = \frac{l}{1 + e \cos(\theta)},$$

where $e = 0$ if the curve is a circle, $0 < e < 1$ if it is an ellipse, $e = 1$ if it is a parabola, and $e > 1$ if it is a hyperbola.

Like many areas of pure mathematics the conic sections were an answer looking for a problem. That problem came with Kepler in the 1610s who was studying the motion of the planets. Kepler took data supplied by the astronomer Tycho Brahe and used it to investigate the latest theories of planetary motion. At that time there were three competing theories of the way that the planets moved in the heavens. The long-standing Ptolemaic theory, in which the Sun and the planets went around the Earth in a combination of circles and epicycles; the recent (and literally revolutionary) Copernican theory in which the planets (including the Earth) orbited the Sun in circles, and theory of Brahe himself in which the planets orbited the Sun, which in turn orbited the Earth. Whilst the Copernican theory had many advantages over the other theories in terms of its simplicity (and extraordinary elegance and power), Kepler found that it

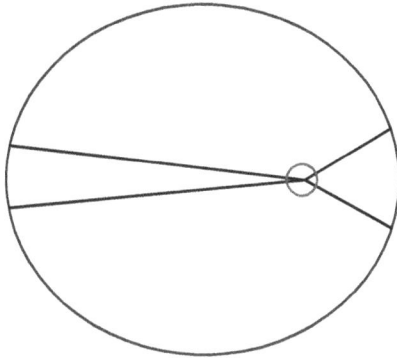

Figure 10.8 *A planet going round the Sun in an ellipse. The regions between the straight lines have equal areas, and are swept out in equal times.*

didn't fit the data especially well, and from the point of view of experimental fit, the Brahe theory was possibly better.

However, Kepler did not abandon Copernicus' theory, instead he realised that it could be improved. The problem he realised was the insistence on circular orbits. The Greeks had chosen circles because they saw them (possibly correctly) as the most perfect of all curves, and thus the only possible orbits of the planets. What other possible orbit could there be? However, Apollonius had worked out the answer 1,800 years before. Kepler realised that if he replaced the circular orbits of the planets by elliptical ones, then everything worked perfectly. This was an astonishing fluke. The laws of motion could have had many solutions, but, to the great fortune of human civilisation, the solution which mattered was one which (in good Blue Peter fashion) someone had made earlier. Kepler went on to formulate his three laws of planetary motion, namely that the planets moved in elliptical orbits with the Sun at the focus, they swept out equal areas in equal times as they went round, and that as the orbital distance cubed, so the orbital period squared. Kepler's first two laws are shown in Fig. 10.8 and we saw Kepler's Third Law in Chapter 1.

These laws allowed very precise calculations of the planetary orbits, which fitted the data perfectly, but no one knew at that stage why they were true. This remained the case until in the late 1600s, Sir Isaac Newton discovered

the laws of motion and the law of gravity. The latter stating that the force acting on a planet from the Sun was inversely proportional to the distance from the Sun squared. Newton then used his newly created theory of calculus (together with a lot of geometry) to prove that Kepler's laws followed directly from his theory of gravitation. Again this was a huge fluke, as most problems in applied mathematics don't have an exact solution even if you can write down the equations describing them. A good example is the laws of fluid motion. However, again luckily for humanity, Newton's equations did have a straightforward solution for the case of a single planet going around the Sun, and this was the ellipse discovered by Kepler. (The law of equal areas corresponds to the conservation of angular momentum, and his third law is a direct consequence of the inverse square law.) As we showed in Chapter 1, so good were Newton's theories that they led to the discovery of the planet of Neptune.

The three-body problem

Such is our faith in Newton's laws that they are now used for large-scale calculations, including such delicate issues as the fate of the Solar System and (indeed) of the whole of humanity itself. This involves calculating the orbit not of a single planet going around the Sun, but of all of the objects in the Solar System. There are a number of problems with doing this, all of which are the subject of significant ongoing research. Firstly, there is the sheer size of the problem, with not only the calculation of the planets but of all of the asteroids and other bodies in the Solar System. Secondly, unlike the case of a single planet and the Sun (the so called *two-body problem*) which (as we have seen) has an exact solution, as soon as we go to three or more bodies, there is simply no closed form solution. To give some idea of the complexity of the problem, the set of Newton's equations for a problem with N planetary bodies of individual mass m_i and at position r_i is given by

$$\frac{d^2 r_i}{dt^2} = -G \sum_{j \neq i}^{N} \frac{m_j \left(r_i - r_j \right)}{\left| r_i - r_j \right|^3}.$$

Thirdly, as well as being essentially impossible to solve analytically, these equations are also hard to solve on a computer. The problem is that we usually want to solve these equations for a long time (to calculate the fate of the Solar System we are talking of billions of years. Over such a long period of time, errors made by the computer in solving the equations can accumulate over time, leading to very inaccurate solutions. However, recently great advances have been made in the development of numerical methods for which the errors made cancel each other out over long times. Such numerical methods, such as the recently studied *symplectic methods* are transforming our understanding of the long-time evolution of the Solar System [61]. The final problem with solving Newton's three-body (or indeed N-body) problem is that the solutions can be *chaotic* in exactly the manner described in Chapter 5. A typical such orbit is shown in Fig. 10.9 and we can see that it is very different from an elliptical orbit.

As we described in Chapter 5, the key fact about a chaotic or bit is that it is highly complex, and very hard to predict even with highly accurate numerical methods. The first person to recognise this was the great French mathematician Henri Poincaré who we also met in Chapter 5. Poincaré who was studying whether the Solar System was stable. (The answer by the way is *maybe*.)

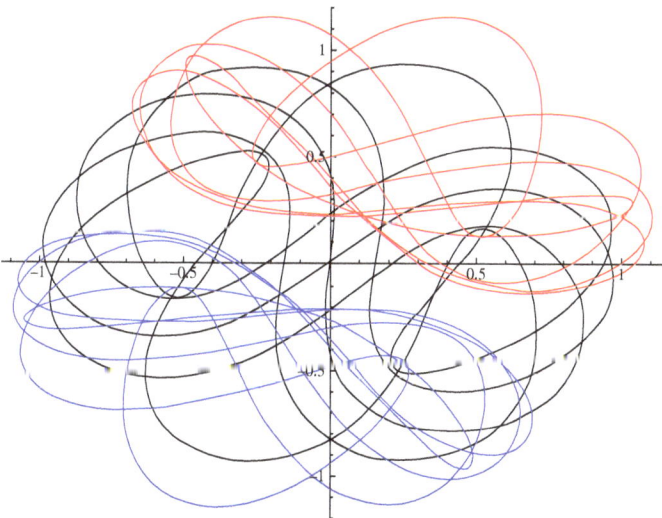

Figure 10.9 *The chaotic trajectory of a solution of the three-body problem.*

Lambert's problem and the elliptical orbit

Whilst, as we have seen, the problem of finding the general orbits of the bodies of the Solar System is hard, the problem of finding the motion of a satellite within this system is fortunately much easier. This is because a satellite is so small that it does not affect the motion of any of the bodies that it comes into contact with. Furthermore, a satellite will mostly be influenced by the gravity of the Sun, unless it comes close to a planet. We will look at this case presently, but will start with the more general case of the satellite moving under the gravity of the Sun. Whilst this is a modern problem for satellites, it was considered 250 years ago by Johann Heinrich Lambert (1728–1777). He was a Swiss mathematician who made a fundamental study into the orbits of bodies in the Solar System using Newton's model of the Solar System, and his work is still in heavy use today in the direction of satellites. In celestial mechanics *Lambert's problem,* which he solved, is the problem of determining the orbit in space which takes a satellite from two different points in a given time. It has important applications in determining the preliminary orbit of the satellite, and allow it to be navigated from one point in the Solar System to another.

Lambert's problem is illustrated by Fig. 10.10 in which the question is to find the trajectory which a body in the Solar System takes if it moves from the point

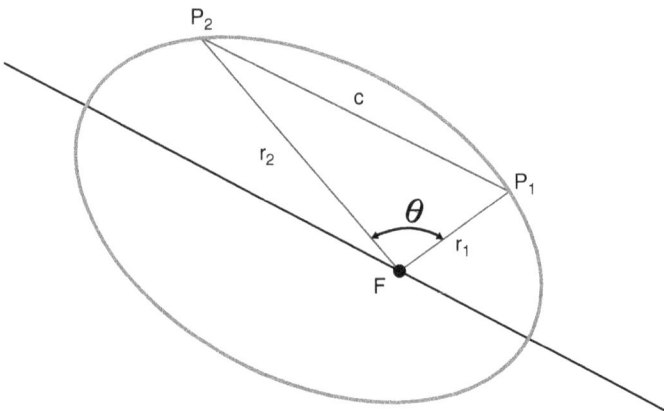

Figure 10.10 *The elliptical orbit of a satellite around the Sun at the focus F of the ellipse.*

P_1 to the point P_2 in a given time T. This problem is solved by assuming that the body moves in an elliptical orbit with the Sun at its focus F (or in the case of a moon shot the Earth will be at the focus), which in turn tells us the angle it has to move in its orbit. The known geometry of the orbit allows the calculation of the precise parameters of the ellipse that it moves on to be made with relative ease. From these the complete trajectory of the satellite can be then calculated with high precision. More details are given in [62]. It is very nice that a model of the Solar System which assumes motion in the classical shape of a conic section discovered over 2,000 years ago, can have such important applications in modern technology.

How some amazing women put a man on the Moon

We now fast forward 200 years to the 1960s in which a pressing need was to calculate the parameters of the orbit of the Apollo space missions to the Moon. This had to be done without the benefits of powerful modern computers. Indeed the on-board computer on the spacecraft themselves had (much) less computing power than that of a modern mobile phone. Instead the orbits were calculated in advance by mathematicians using versions of the Lambert problem described above, together with a lot of other clever calculation methods (often done using a slide rule or a mechanical calculator). Remarkably (for the time) three of these were African American women including the mathematician Katherine Goble, the engineer Mary Jackson, and their supervisor Dorothy Vaughan. Their work is celebrated in the Hollywood film *Hidden Figures* (look out for the giant chalkboard) and in the book [63]. Their calculations allowed manned spacecraft to be accurately steered into a low Earth orbit, to then be 'injected' onto a path to the Moon, and then 'inserted' into a Lunar orbit, as illustrated in Fig. 10.11.

So far so good, but in the case of Apollo 13 in 1970 the calculations had to be rapidly revised when the spacecraft was severely damaged on the way to the Moon by the explosion of one of its Liquid Oxygen tanks followed by the shutting down of the Command Module. Following this the astronauts had

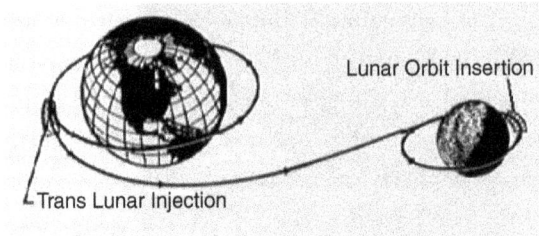

Figure 10.11 *A trajectory of a Moon mission, starting from an orbit around the Earth, followed from a passage from the Earth to the Moon, and then an orbit around the Moon.*

to enter the Lunar Module which had to then take them back to Earth along an unplanned orbit. It is a huge tribute to the orbit calculators at NASA in Houston, working under extreme pressure, that the astronauts were able to return safely to Earth. All of this happened when I was 10. My teacher at primary school involved us all thoroughly in the unfolding drama of this mission, and we all rejoiced in the safe return of the astronauts. This made a huge impression on me at the time and was one of the reasons that I got thoroughly switched on to mathematics and science. A great account of this is given in *Lost Moon* [64].

Sling shots and hyperbolic orbits

Whilst the orbits of bodies around the Sun are ellipses, it is another type of orbit, the hyperbolic orbit, which plays a vital role in long distance travel in the Solar System. Suppose that you are on the surface of a body, such as the Earth, and throw a projectile into space with a velocity v. Our usual experience is that the projectile will slow down due to the effects of gravity, and will eventually reverse its direction before falling back to the Earth. However, if you throw the particle with a high enough velocity v then it will 'escape the gravity of the Earth' and will continue to move to infinity (or at least as far as the edge of the universe). For the Earth this escape velocity V_e is given by

$$V_e = \sqrt{2gR} = 11.2 \text{ km s}^{-1}.$$

This value is 41% higher than the velocity needed to insert a satellite into a near Earth orbit. (On the event horizon of a black hole this velocity is the speed of light of 3 million km/s.) If the velocity v is higher than the escape V_e then the body will move on a hyperbolic orbit around the Earth. Similarly if a body moves faster than the escape velocity of the Sun then it will move on a hyperbolic orbit around the Sun, as illustrated in Fig. 10.12.

Roughly speaking a hyperbolic orbit comprises an almost straight section which approaches one of the asymptotes of the hyperbola, along which the body has a near constant velocity of approach relative to the Sun which is v_∞ and is given by the formula

$$v_\infty^2 = v^2 - V_e^2.$$

This is called the *hyperbolic excess velocity*. The body then speeds up as it approaches the Sun (or any other large object) reaching a maximum velocity of v close to the Sun. It then swings around the Sun and leaves on another straight path, approaching the other asymptote of the hyperbola, reaching a departing velocity of v_∞. By doing so it changes its direction of motion.

This phenomenon is exploited in the *slingshot effect* which is used to take satellites to the distant planets by swinging them around other large planets

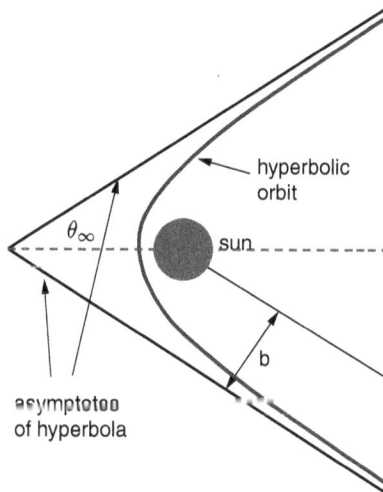

Figure 10.12   *A hyperbolic orbit around the Sun, showing the limiting asymptotes and the deflection angle θ_∞.*

on the way. The slingshot effect is used to change their direction, and the gravitational pull of the large planet is used to speed them along their way. Consider for example a satellite going to Pluto via the large planet Jupiter. If guided correctly it will go on a hyperbolic orbit around Jupiter approaching on one asymptote of the hyperbola. As it approaches Jupiter at a relative approach velocity of v_∞ it will accelerate towards it because of its gravity attraction, swing around it on a hyperbolic orbit and then move away at a departing velocity of v_∞ (again) relative to Jupiter and along the other asymptote. The angle of its path has changed. However, whilst all of this is happening Jupiter is moving around the Sun at an orbital velocity of about 13 km s^{-1}. The effect of being dragged along by Jupiter not only changes the angle of its path but can also increase its speed significantly without any fuel being burnt. Essentially the gravitational attraction of Jupiter gives the satellite some additional energy to continue its orbit. As no energy can be gained or lost in the encounter the same amount of energy is lost by Jupiter, but as it is so massive its orbit is scarcely affected.

This can all be made very precise mathematically by using the known geometry of the hyperbolic orbit. The key parameters of a sling shot around a massive planet of mass M are the approach velocity v_∞ of the satellite relative to the planet and the (so called) *impact parameter b* (illustrated in Fig. 10.12) which it the closest approach that the satellite would make to the planet if it were not affected by the gravity of the large planet. It then follows from the Newtonian model of the Solar System that the satellite's orbit is deflected by an angle $2\pi - 2\theta_\infty$ between the lines of approach and departure where

$$\tan\left(\theta_\infty\right) = \frac{bv_\infty^2}{\mu}.$$

Here as before we have

$$\mu = GM,$$

where G is the Gravitational constant and M is the mass of the large planet. The larger that M is, the more the satellite will be deflected by the slingshot effect.

In a deep space mission the designers of the trajectory of the satellite may make use of a number of gravity-assisted slingshots to propel to satellite to the distant planets. In Fig. 10.13 you can see the orbits of the Voyager probes (launched in 1977) as they passed by Jupiter and Saturn on their way to Uranus, Neptune and the edges of the Solar System and beyond (and according to the film *Star Trek: The Motion Picture*, back again).

Similarly, the Galileo probe to Jupiter was launched from the Earth, had a gravity assisted sling shot from Venus, returned back to Earth twice to have further gravity assists all of which raised its orbital energy. Finally it made it to Jupiter after six years. All it needed to do this was a small excess velocity of 3 km/s above the escape velocity needed to leave the Earth's orbit. This compares with the much shorter, but much more expensive (in fuel), method of launching

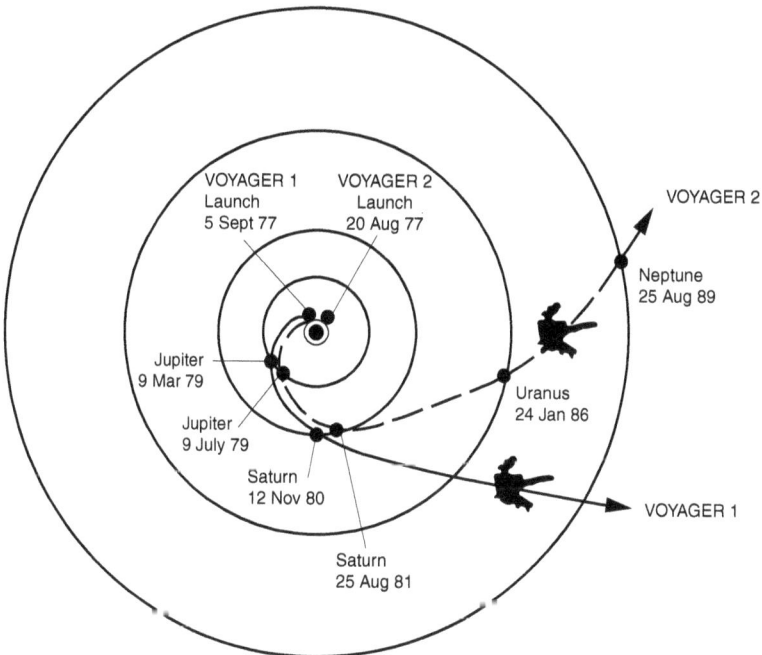

Figure 10.13   *The trajectory of the Voyager satellites, which used the slingshot effect around Jupiter and Saturn to reach Uranus and Neptune.*

the satellite directly at Jupiter. A film of the projected orbit of the European Space Agency JUICE satellite to Jupiter is given in [65].

5. To deep space and back again

As we have seen above, Newton's laws of motion do a remarkably good job in predicting the motions of satellites and the planets. If we think of the modelling process, and **Step 7** then any model which used Newton's laws was felt to be complete by any astronomer up to the start of the 20th Century (and indeed as we have seen above, they still work incredibly well in the Solar System). The physicists of the 19th Century were confident that the laws of physics had all been determined, and that only a few constants needed to be found to complete the models that they used to describe the universe. They were wrong! In 1905, possibly one of the most remarkable years in the history of science, Einstein published three astonishing papers which transformed science and the way that we look at the world. These were on Brownian Motion and Kinetic Theory, the Photoelectric Effect in Quantum Theory and (possibly most famously) his paper on the *Special Theory of Relativity (SR)*.

In 1915 Einstein followed this with the publication of his wholly remarkable paper on the *General Theory of Relativity (GR)*. This latter paper developed a model for the Universe which departed from Newtonian mechanics and gave a (new) theory of gravity. This explained gravity in terms of the distortion of space-time by massive bodies. (Note that for speeds rather lower than the speed of light, and for bodies of the mass of the Earth and the Sun, then the equations of General Relativity very closely approximate those of Newtonian mechanics. This is why the model of the Universe based on Newtonian Mechanics is still very useful for most calculations.) The General Theory of Relativity is summarised by the *Einstein Field Equations* which represent how the *curvature G* of space time is changed by the *mass tensor T* and have the form

$$G_{ab} = 4\pi T_{ab}.$$

Although these may look simple, they are a shorthand for a large number of simultaneous partial differential equations, and are very hard to solve. It is fitting that they should be hard to solve as these equations *describe the whole universe.*

Since their publication in 1915, many mathematicians (including Steven Hawking and Roger Penrose) have worked very hard to find solutions to these very challenging equation. These mathematical studies have led to a number of extraordinary predictions from the General Theory of Relativity. One of these was that light should be deflected in a certain way by a massive body (such as the Sun). This prediction was validated by Sir Arthur Eddington in a famous experiment in 1919 when he made measurements of the positions of stars behind the Sun during a total eclipse. His measurements exactly agreed with Einstein's theory giving confidence in its predictive powers which went beyond, and differed from, the predictions of Newtonian mechanics. A second prediction was that the orbit of the planet Mercury should slightly differ from the true ellipse predicted by Newtonian mechanics, with the ellipse advancing by 23 seconds of arc every Mercurian year. We saw in Chapter 1 that this deviation from a true ellipse had been known about for some time and that existence of the new planet Vulcan had been postulated to explain it. Einstein showed that no such planet was necessary.

Two more predictions were the existence of Black Holes (massive stars for which the escape velocity V_e is higher than the speed of light), and of the expansion of the Universe. Although it took longer to find, there is now extremely strong evidence to support both predictions. Indeed massive Black Holes are now known to be at the centre of galaxies such as our own. One year after the publication of his theory, Einstein made another mathematical prediction, that of Gravitational Waves. These are ripples in space time which travel at the speed of light. This prediction has taken much longer to verify. This is because, despite being created by huge astronomical events, such as the collision of Black Holes or of neutron stars, by the time that they reach us, the ripples through space-time are less than the width of an atom. However, in a piece of work that has recently (and very correctly) been awarded the 2107 Nobel Prize for physics (to Rainer Weiss, Barry Barish and Kip Thorne), Gravitational Waves have been detected. This discovery was made on the

14 September 2015, by the bar detectors of LIGO (Laser Interferometry Gravitational Wave Observatory) detector in Louisiana which measured the Gravitational Waves given off by the collision of two massive Black Holes a billion light years away. Since 2015 there have been several recordings of gravitational waves, with three coming from the collision of Black Holes, and very recently one from the collision of two neutron stars. A whole new era of astronomy using Gravitational Waves has just started. This is a splendid demonstration that a mathematical prediction can come true and have profound consequences.

We now come back to Earth, or at least close to Earth, with a wonderful practical application of the General Theory of Relativity to satellite technology. One of the most important uses of modern satellites is in GPS navigation. Most of us will have GPS location devices in our car and/or our smartphones. It is essential for the accuracy of GPS systems for the satellites to tell the time to very high precision, as the navigational method relies on measuring the time difference between the signals received on the Earth from a number of GPS satellites. If Newtonian mechanics were completely accurate then a satellite would tell the same time as on the Earth. However, owing to the effects predicted by both the special and general theories of relativity this is not the case. Firstly the satellites are travelling at a high speed which causes their clocks to run slow (a prediction of special relativity) by about 7 microseconds per day. Secondly the Earth's gravity is weaker at the satellite than on the surface of the Earth. According to the General Theory of Relativity this also causes the clocks to run faster, by about 45 microseconds per day. The total correction due to relativistic effects is then 38 microseconds per day. This leads to an improvement in GPS accuracy of the order of hundreds of metres. This can make all the difference there is to the use of the GPS algorithms in applications such as navigation, surveying, or landing an aircraft.

6. A bit of origami

We finish this chapter by looking at a link between space, maths and the art of origami (which we have already touched on in Chapter 9). When a satellite

Figure 10.14 *Unveiling the photo cells of a satellite, which were folded up by using origami. (Credit: With thanks to Larry Howell, Robert Lang and Spencer Magleby)*

is sent up into space it has to fit inside a small rocket. However, when it gets into space then it has to deploy large solar panels in order to gain energy from the Sun. The designers of the satellites thus have the problem of how to fold a solar panel into a small space. Fortunately mathematicians already have the answer to this problem in the shape of the mathematical algorithms now used to design origami patterns. One of the leading figures behind this is Robert Lang, who we also met in Chapter 9, and is simultaneously a mathematician, an origami master and a rocket scientist. Figure 10.14 shows how mathematical origami is used to unfold the photo cells of a satellite. Who could ask for a better application of mathematics. You can find out more about the uses of origami in space science in the video in [66].

7. The end of this book, and the beginning of the next

This chapter brings us to the end of this book and of the examples of the use of mathematical modelling. We have now seen the many ways that mathematical models impact on many areas of our lives, from modern technology to combatting diseases, and from saving the whales to predicting the climate. In this chapter we have even seen a mathematical model (Einstein's General Theory of Relativity) of the whole Universe. We have also seen that in order to solve most mathematical models, and to make useful predictions, it is essential to combine mathematical reasoning with the power of the modern computer. This helps us to develop algorithms (such as a modern weather forecast, or the computer programs used to control the National Grid) which can both help us to predict the future and also to keep control of the present. All of these algorithms, and predictions, have to be checked and validated by comparing them to data. The next logical step is to develop algorithms based just upon the data alone. These lie at the heart of the rapidly growing area of *machine learning*, which is challenging, and greatly expanding, the way that we model the real world. This subject is worth a whole book in itself. Watch this space!

References

[1] I. Asimov, (1964), *Asimov's Biographical Encyclopedia of Science and Technology*. Doubleday.

[2] T. Sims *et al.*, (2007), *101 Things to Do During a Dull Sermon*. Monarch Books.

[3] G. Box, (1976), "All models are wrong but some are useful", *Journal of American Statistical Association*. A discussion of this comment is given in https://en. wikipedia.org/wiki/All_models_are_wrong

[4] J. D. Murray, (1989), *Mathematical Biology*. Springer.

[5] H. Fry, (2015), *The Mathematics of Love: Patterns, Proofs, and the Search for the Ultimate Equation*. TED Books.

[6] M. Barons *et al.*, (2019), "Study groups with industry: What is the value", *Mathematics Today*.

[7] P. Wilmott, (2019), *Machine Learning: An Applied Mathematics Introduction*, Panda Ohana Publishing.

[8] S. Turnnidge, (2021), "How media reporting of modelling has shaped our understanding of the pandemic", *Full Fact*, https://fullfact.org/health/scientific-modelling-covid/

[9] Laplace's Demon, https://en.wikipedia.org/wiki/Laplace%27s_demon

[10] E. N. Lorenz, (1963), "Deterministic nonperiodic flow", *Journal of the Atmospheric Sciences*, Vol. 20, pp. 130–141.

[11] *The Butterfly Effect*, (2004), Film directed by Eric Bress and J Mackye Gruber.

[12] R. Bradbury, (1953), "The sound of thunder", *The Golden Apples of the Sun*. Doubleday.

[13] J. Gleick, (1987), *Chaos: Making a New Science*. Viking Penguin.

[14] I. Stewart, (1989), *Does God Play Dice? The Mathematics of Chaos*. Wiley-Blackwell.

[15] S. Strogatz, (1994), *Nonlinear Dynamics and Chaos*. Addison-Wesley.

[16] P. G. Drazin, (1992), *Nonlinear Systems*. Cambridge University Press.

[17] M. Gladwell, (2000), *The Tipping Point: How Little Things Can Make a Big Difference*. Little, Brown and Company.

[18] K. Falconer, (2003), *Fractal Geometry: Mathematical Foundations and Applications*. John Wiley & Sons.

[19] D. Acheson (1997), *From Calculus to Chaos: An Introduction to Dynamics*. Oxford University Press.

[20] Recent IPCC reports are given in https://archive.ipcc.ch

[21] The 2016 Paris Agreement is described in https://en.wikipedia.org/wiki/Paris_ Agreement

[22] HRH The Prince of Wales, Tony Juniper and Emily Shuckburgh, (2017), *Climate Change*. Ladybird Expert Books Ltd., London.

[23] The 6th IPCC assessment report (for 2021) is given in https://www.ipcc.ch/report/sixth-assessment-report-working-group-i/

[24] The website http://wxcharts.eu gives an excellent set of easy to follow, and up-to-date, weather charts.

[25] C. Budd, M. Cullen and C. Piccolo, (2016), "Improving weather forecasting accuracy by using r-adaptive methods coupled to data assimilation algorithms", in P. J. Aston *et al.* (eds.), *UK Success Stories in Industrial Mathematics*. Springer International.

[26] J. Imbrie and K. P. Imbrie, (1979), *Ice Ages: Solving the Mystery*. Macmillan.

[27] H. Kaper and H. Engler, (2013), *Mathematics and Climate*. Society for Industrial and Applied Mathematics.

[28] T. Lenton, (2016), *Earth System Science: A Very Short Introduction*. Oxford University Press.

[29] D. MacKay (2009), *Sustainable Energy without the Hot Air*, UIT Cambridge Ltd. https://www.withouthotair.com/download.html

[30] C. Budd and S. Morupisi, (2021), "An analysis of the periodically forced PP04 climate model, using the theory of non-smooth dynamical systems", *IMA Journal of Applied Mathematics*, pp. 76–120.

[31] C. Budd and C. Sangwin, (2004), "101 uses of a quadratic equation", *Plus Magazine*, https://plus.maths.org/content/101-uses-quadratic-equation

[32] F. Li, "Demand side response, conflict between supply and network driven optimisation", Poyry, University of Bath report to DECC, November 2010.

[33] C. Budd and J. Wilson, (2002), "Bogdanov-Takens bifurcation points and Silnikov homoclinicity in a simple power-system model of voltage collapse", *IEEE Transactions on Circuits and Systems I: Fundamental Theory and Applications*, Vol. 49, pp. 575–590.

[34] For details of the SMART Grid, see https://en.wikipedia.org/wiki/Smart_grid

[35] The NE US blackout is described in the website https://en.wikipedia.org/wiki/Northeast_blackout_of_2003

[36] The World Cup semi-final match close run for the National Grid, is described in the article http://news.bbc.co.uk/1/hi/uk/5059904.stm

[37] C. Skittides and W-G. Fruh, "Wind forecasting using Principle Component Analysis", https://www.researchgate.net/publication/261759110_Wind_forecasting_using_Principal_Component_Analysis

[38] P. J. Nahin, (2006), *Dr. Euler's Fabulous Formula: Kills Many Mathematical Ills*. Princeton University Press.

[39] R. Wilson, (2018), *Euler's Pioneering Equation: The Most Beautiful Theorem in Mathematics*. Oxford University Press.

[40] D. Luenberger, (2008), *Linear and Nonlinear Programming*. Springer.

[41] The Wikipedia article on the Simplex Algorithm is given by: https://en.wikipedia.org/wiki/Simplex_algorithm

[42] L. R. M. Wilson, N. C. Cryer and E. Haughey (2019), "Simulation of the effect of rainfall on farm-level cocoa yield using a delayed differential equation model", *Scientia Horticulturae*, Vol. 253, pp. 371–375.

[43] D'Arcy Thompson, (1917), *On Growth and Form*. Cambridge University Press.

[44] https://en.wikipedia.org/wiki/Population_dynamics_of_fisheries

[45] http://news.bbc.co.uk/1/hi/sci/tech/6592693.stm

[46] The work of the Limerick based MACSI group on Guinness is described in the article http://ulsites.ul.ie/macsi/sites/default/files/macsi_the_initiation_of_guinness.pdf

[47] S. Howison, (2005), *Practical Applied Mathematics: Modelling, Analysis, Approximation*. Cambridge University Press.

[48] C. Budd and A. Hill, (2011), "A comparison of models and methods for simulating the microwave heating of moist foodstuffs", *International Journal of Heat and Mass Transfer*, Vol. 31, pp. 807–817.

[49] M. Miodownik, (2014), *Stuff Matters: The Strange Stories of the Marvellous Materials that Shape Our Man-made World*. Penguin.

[50] J. Cosgrove and M. Ameen, (1999), *Forced folding and Fracture*. Geological Society of London.

[51] R. de la Rue and S. de la Rue, (2008), "Introduction to photonic crystals and photonic band-gaps", *Photonic Crystals: Physics and Technology*, pp. 7–25.

[52] R. A. Bagnold, (1941), *The Physics of Blown Sand and Desert Dunes*. London: Methuen.

[53] A nice description of the optics behind the Harry Potter cloak can be found in the article http://www.independent.co.uk/news/science/invisibility-cloak-rochester-cloak-digital-harry-potter-a7039106.html

[54] N. J. McCullen, D. P. Almond, C. J. Budd and G. W. Hunt (2009), "The robustness of the emergent scaling property of random RC network models of complex materials", *Journal of Physics D: Applied Physics*, Volume 42.

[55] R. Lang, (2011), *Origami Design Secrets: Mathematical Methods for an Ancient Art*. CRC Press.

[56] R. Thomas, (2014), "Trisecting an angle with origami", *Plus Magazine*, https://plus.maths.org/content/trisecting-angle-origami

[57] *Maths Today* which is the popular journal of the Institute of Mathematics and its Applications has a special edition on *Maths in Space* (October, 2017) which contains lots of information relevant to this chapter.

[58] C. F. Gauss, (1801), *Disquisitiones Arithmeticae*.

[59] C. Budd and C. Sangwin, (2000), "Analemmatic sundials: How to build one and why they work and how they work". *Plus Magazine*, https://plus.maths.org/content/analemmatic-sundials-how-build-one-and-why-they-work

[60] A. C. Clarke, (1945), "Extra-terrestrial relays", *Wireless World*, Vol. 11, No. 10, pp. 305–308.

[61] E. Hairer, C. Lubich and G. Wanner, (2002), "Symplectic integration of Hamiltonian systems", *Geometric Numerical Integration*, pp. 167–208.

[62] Orbital mechanics, http://www.braeunig.us/space/orbmech.htm

[63] The story of the amazing women who worked for the Apollo Space Program is described in the book *Hidden Figures* by Margot Lee Shetterly, (2016), published by William Morrow and Company.

[64] J. Kluger and J. Lovell, (1994), *Lost Moon: The Perilous Voyage of Apollo 13*. Pocket Books.

[65] An animation for the orbit of the ESA Juice probe to Jupiter is given in https://www.youtube.com/watch?v=uMyaIphWp1A

[66] Origami in space, https://www.youtube.com/watch?v=3E12uju1vgQ

[67] I. Newton, (1687), *Principia Mathematica*.

Index

A

A0, 240–241

A4, 242, 244

abstract mathematics, 155

acceleration, 9

Adams, Douglas, 27

Adams, John Couch, 19

ad-hominem attacks, 138

air changes, 122

aircraft, 22

air resistance, 8

albedo, 129, 139–140, 142

algebra, 30

alloys, 229–231

alternating current (AC), 148,
    152–154, 157, 169

alternating voltage, 157

Aluminium Oxide, 236

amplitude, 153

analemma, 253

analemmatic sundial, 253–254

analogue computer, 251

analysis, 16

angular momentum, 262

angular velocity, 257

annealed steel, 230

Antarctic, 118

Antarctic ice, 140

antenna, 45

Apollo, 250, 259, 265

Apollo 13, 265

Apollonius, 259–261

applied mathematics, 181, 217

arbitrary constant, 53

Arctic, 109

Arctic ice, 135

Arctic permafrost, 134

Arctic sea ice, 110–111

Arranz, Guillermo Jimenez, 45

Arrhenius, Svante, 106

Asimov, Isaac, 6

asteroid, 64

astrology, 249

asymptotes, 267

asymptotic methods, 27

Atlantic circulation, 143

Atlantic Ocean, 109, 135

atmosphere, 106, 121, 130–131, 133,
    141

auto regressive moving average, 173

B

Babasola, Tosin, 180

Babylonian, 243, 250–251

Babylonian cuneiform, 164

Barish, Barry, 271

Barnsley Fractal, 98

Bath, 105, 222

BBC data, 57

beam, 210, 214–216

beer, 177

bell curve, 113

bending equation, 214

bifurcation, 92

bifurcation diagram, 92–93

biological sciences, 22–23

biological systems, 22

biology, 63

bitter, 191–192, 206–207

black body, 131

black body radiation, 130

black hole, 20, 266, 271–272

blackout, 163

black swan event, 63, 96

Blondel, Philippe, 38, 45
Blue Peter, 261
Bohr, Niels, 63
Book of Genesis, 250
Box, George, 15
Box Models, 128
Bradbury, Ray, 80
Brahe, Tycho, 260–261
Brahmagupta, 164
brain, 22
Brazil Nut Effect, 232–233
breakfast, 232
Bristol, 79–80
Bristol Zoo Gardens, 194
British Science Festival, 189–190
Brownian Motion, 270
bubbles, 190–193
Bubonic Plague, 59
buckle, 218
bucky-ball, 228–229
Bude, 220–221
bull moose fold, 245
bus, 158, 160
butterfly effect, 80, 84–85
butterfly shaped strange attractor, 85

C
Cahn–Hilliard equation, 230–231, 237
calculus, 9, 16, 76, 146, 156, 249
calendar, 250
calibration, 218
camel, 94
camel's back, 93–94
cantilevered beam, 215
capacitor, 238–239

Carbon Capture Technology, 143
Carbon Dioxide, 106–107, 114–116, 119–120, 124, 127, 130, 133–138, 141–143, 168, 191, 193
carbon fibre, 236
carbon technology, 167
card, 213–214
cardboard, 216–217, 244
Cartesian coordinates, 259
Cartwright, Dame Mary, 81, 86
catastrophe theory, 93
cavity, 200
cellular structures, 192
centripetal acceleration, 256
chandelier, 6, 9, 75
chandelier pendulum, 7
chaos theory, 64, 68–70, 82–86, 91, 96–101, 104, 138
chaotic, 80, 84, 123, 263
chaotic behaviour, 79, 81, 90, 92
chaotic billiards, 81
chaotic motion, 74–75, 78, 80
chaotic solutions, 138, 169
chaotic system, 80, 85
chaotic trajectory, 263
chaotic water pendulum, 79
chevrons, 222, 224
chicken, 193
Chinese, 239, 250
Chinese Remainder Theorem, 251
Chipping Campden Food Research Association (CCFRA), 199
chocolate, 177
Christiaan Huygens, 7
circle, 259–260
Clarke, Arthur C., 257

Clarke Belt, 258
Clarke Orbit, 258
classical rocket equation, 258
climate, 2, 5, 62, 69, 101, 121, 124,
   127, 180, 238, 274
climate change, 3, 25, 96, 104, 106,
   108–128
climate forecasting, 123–124
climate model, 21, 120–126, 129,
   135–139
cloaking device, 233
cloak of invisibility, 233–235
clockwise, 253
clockwork universe, 68
coastal flooding, 112
cocoa, 178, 180
cocoa growing, 24
Cocoa Research Institute of Nigeria,
   180
coffee, 193
collisions, 38
complex, 160, 164, 220
complexity, 235
complex material, 236
complex number, 155–158, 219
complex voltage, 158, 160
composite material, 236
computer, 11–12
computer algorithm, 11, 17
computer graphics, 97
computer models, 62
concrete, 225
conductance, 160, 238
conductivity, 239
congruences, 251
conic sections, 259–260

constrained optimisation problem, 178
constraints, 179
continental drift, 220
continued fraction, 243
convection, 83
cook, 197
Copernican theory, 260
Copernicus, 261
Cornwall, 220–221, 224–225
Council of Nicaea, 251
COVID-19, 2, 7, 14, 26–27, 48–50,
   52, 55–56, 58–59, 63–64, 84
crops, 181
crowds, 238
crystals, 225, 227–229
current, 153–154

D

Dantzig, 180
data assimilation, 126
da Vinci, Leonardo, 13, 214
de Fermat, Pierre, 53
deflection, 214
deformation, 214
degrees of freedom, 74
de Havilland Mosquito, 235
delay, 30, 33, 35
Delta-variant, 55, 60
Demaine, Erik, 245
Department of Energy and Climate
   Change, 167
differential equation, 2, 7–10, 20, 27,
   51–52, 54, 65, 76–77, 169, 181,
   214
differentiation, 156
digital watches, 225

dinosaurs, 63
direct current (DC), 148, 152–155, 169
direct pulse, 43
discretisation, 122–124
diseases, 2
double pendulum, 71–82, 88
drink, 176
dynamical system, 87, 93–94, 104,
   164, 166

E
Earth, 117, 122, 131–134, 138–139,
   141–142, 211, 248, 250–253, 255,
   257, 270
Earth Intermediate Complexity models,
   128
earthquake, 211–212
Earth's atmosphere, 254
Earth's axis, 250
Earth's rotation, 255–256
Earth System Model, 105
Easter Sunday, 251
echo, 42
echo pulse, 43
eclipse, 271
economics, 24
Eddington, Arthur, 271
Ed Hawkins, 127
Edison, 146, 154
EEG, 70
egg, 193–196
Egyptian, 251
Eifel, 215
eight great technologies, 177, 249
Einstein, Albert, 16, 20, 270–271
Einstein Field Equations, 270

Einstein's General Theory of Relativity,
   20, 274
elastic, 214
elasticity, 189
elastic material, 213
elastic medium, 211
electrical field, 65, 202, 235
electrical supply networks, 237
electrical waves, 236
electricity, 2, 146–152, 161
electricity supply, 169, 172
electromagnetic field, 199
ellipse, 253, 259–262, 264, 266, 271
elliptical, 18, 263
elliptical orbit, 257, 264–265
elliptic integrals, 10
El Nino, 108, 116, 135
El Nino Southern Oscillation (ENSO),
   68–69
emerge, 236
emergent properties, 235
emissivity, 132, 134
energy, 146, 229
energy balance, 132
energy balance equation, 142
energy balance model (EBM),
   128–129, 133, 138–139
energy budget equation, 133, 138
energy budget formula, 135
energy supply networks, 173
enthalpy, 186
enthalpy equation, 186–188, 203
environment, 2, 104
environmental change, 106
European Projects for Ice Coring in
   Antarctica (EPICA), 118–119

epicycles, 260
epidemic, 2–3, 49–55
equation of time, 253
ergodic properties, 100–101
escape velocity, 271
Euler, Leonhard, 12, 157, 214
Euler's theorem, 157
evolutionary, 63
Excel, 88
exponential, 88
exponential function, 53
exponential growth, 55
Extra High Voltage (EHV), 168
extreme events, 170

F
4-cycle, 89
Faraday, Michael, 20, 65, 146
farm, 177
Fast Fourier Transform (FFT),
    185
feasible region, 179
fermentation, 177
fibre-optic cable, 225–227
final equation, 122
finite difference, 196
finite element, 196
first derivative, 9
fish, 40, 177, 182, 188–189
fishing, 181
five-day forecast, 173
fixed points, 94–95, 164
flow, 35
fluid mechanics, 182
foam, 190–192
focus, 265
fold bifurcation, 95

folding pattern, 221, 244–245
folding rocks, 2
food, 2, 176, 232
football matches, 48
force, 122, 211
formula, 27
fossil fuels, 124
Foundation Series, 24
Fourier Analysis, 185
Fourier Coefficient, 184–185
Fourier, Joseph, 183
Fourier series, 184, 187, 196
Fourier's heat equation, 202
fourth order, 215
fourth-order differential equation, 216,
    218–219
fractal, 97–99
Fractal, Barnsley Fern, 97
fractal form, 98
fractal set, 86
freezer lorry, 205
freezing, 182–190
frequency, 40, 66, 157, 225–226, 236,
    244
frequency Wi-Fi, 81
freshness, 182–190
friction, 8
Financial Times Stock Exchange
    (FTSE), 68–69
future, 62

G
Galilei, Galileo, 5–7, 12–14, 71, 214,
    256
Galle, Johan, 19
gambling, 62
Gascoigne, Paul, 151

general circulation model (GCM), 125, 128, 136
General Theory of Relativity, 270–272, 274
geology, 63
geometrical series, 240
geometry, 27, 234
geo-sciences, 219
geostationary orbit, 257
geostationary satellites, 257–258
geosynchronous orbit, 258
glacial cycle, 118, 136
Gleick, James, 86
global warming, 106–107, 115
gnomon, 252–253
Goble, Katherine, 265
God, 250
gold, 224
Google, 73
Gosset, William, 189–190
government, 48
GPS, 249, 258, 272
grain silos, 232
granular material, 232–233
gravitation, 17, 262
gravitational acceleration, 257
gravitational constant, 257, 268
gravitational force, 8
gravitational waves, 271–272
gravity, 133, 262, 266
Great Storm of 1987, 63
greenhouse effect, 106, 115, 136
greenhouse gases, 133
greenhouse gas methane, 143
Greenwich Mean Time (GMT), 253
Gregorian calendar, 251

Gresham College, 211–212
grid, 147, 162, 172
Guinness, 189–190, 192–193

H
HADGEM3, 125
Hadley Centre, 108, 125–127
Harry Potter, 211, 233–235
Hawking, Steven, 271
heat equation, 183–186, 195
Heinlein, Robert, 124
Helmholtz equation, 226
herd immunity, 58
hermodynamics, 121, 182
Herschel, William, 19
Hertz, Heinrich, 20
Hidden Figures, 265
high voltage, 149
Hill, Andrew, 202
hockey stick, 118
Hooke, Robert, 211–212
Hooke's law, 211
horizontal orbit, 255
House of Commons, 162
Houses of Parliament, 253–254
Hull, Tom, 245
human behaviour, 49
Hunt, Giles, 222
hydropower, 146
hyperbola, 259–260, 267–268
hyperbolic excess velocity, 267
hyperbolic orbit, 266–270

I
ice, 109–112, 123, 137, 139
ice age, 118–119, 135–137
ice-albedo effect, 139

ice-albedo feedback, 142
ice cores, 117
imaginary number, 155–156, 161, 220, 238
impact parameter, 268
incubation, 194
incubators, 193
inductance, 161
infected population, 50
infectious disease, 50
infra-red radiation, 131
in phase, 72–73, 77–78
insert velocity, 256–257
insolation, 116
integration, 156
Intergovernmental Panel for Climate Change (IPCC), 104, 113, 115, 117, 127–128, 139
inverse problem, 212
invisibility cloak, 211, 235
irrational number, 243
irreversible change, 94
ivory tower, 26

J
Jackson, Mary, 265
Julian calendars, 251
Jupiter, 268–269

K
Keeling Curve, 114
Kelvin, Lord, 107, 141, 182
Kepler, 259–262
Kepler's laws, 262
Kepler's Second Law, 18
Kepler's Third Law, 18

kilo Watt hour (kWh), 147
Kinetic Theory, 270
kink bands, 222
knapsack problem, 206
Korolev, Sergei, 255

L
lager, 191, 206
Lagrange, 75
Lagrangian, 75
Lambert, Johann Heinrich, 264
Lambert's Law, 200–201
Lambert's problem, 264–265
Lang, Robert, 245, 273
Laplace, Pierre Simon, 66–67, 70, 75
Laplace's Demon, 68, 80
Laser Interferometry Gravitational Wave Observatory (LIGO), 272
latent heat, 186
latitude, 252
laws of motion, 74
laws of physics, 121
layer, 214
LCD screens, 235
leap year, 251
Lebesgue, 75
Legendre, 75
Leibniz, Gottfried, 76
Lenton, Tim, 143
Le Verrier, Urbain, 19
Limerick, 193
linear, 65, 179
linear constrained optimisation problem, 179
linear differential equation, 77

linear equation, 219
linear problems, 180
linear programming problem, 178
linear system, 77
liquid, 213
liquid crystals, 210
Little Ice Age, 116, 118, 138
lockdown, 59
logarithm, 239
logarithmic mixing, 238
logistic equation, 89–91, 181
logistic map, 86–93, 97, 100–101
long-wave heating, 131
long wavelength radiation, 130
long-wave radiation, 131, 133–134
Lorenz, Edward Norton, 80, 82–84
Lorenz equations, 83–85, 88, 93
Lunar Module, 266

M
machine learning, 173, 274
Mackay, David, 143
magnetic field, 65, 235
magnetron, 198–199, 202
Malthus, Thomas, 87–88, 181
Mandelbrot set, 97–98
map, 88
March equinox, 251
Mars, 249
marshmallow, 200
mass of the Earth, 257
mass tensor, 270
materials, 2, 101, 189, 210–211, 225, 229
mathematical algorithms, 180
mathematical drunkard, 25, 77

mathematical modelling, 2–3, 7, 9–10, 12, 14–17, 24–27, 30, 40–46, 48–49, 62, 64, 68, 71, 77, 115, 121, 128–137, 148, 154, 161, 163, 171, 176, 180, 183, 194, 197, 199, 213, 219, 237, 248–250, 273–274
Maths Today, 249
matrix, 238
mattress, 217–218
Mauna Loa, 114
Maxwell, James Clerk, 65–66, 68
Maxwell's equations, 82
Maxwell's laws for electromagnetism, 65
measles, 51
mechanical clock, 253
Medley, Graham, 60
Mercury, 19–20, 271
mergent, 236
meta-materials, 235
methane, 130, 134
Met Office, 105, 108, 122, 125–126, 128
microwave, 197, 199, 201–204
microwave cooker, 24, 66, 197–205
microwave field, 200–201
microwave oven, 200
Mid Pleistocene Transition, 136
Milankovitch, 116
Milankovitch cycles, 136, 138
Millock Haven, 224
mission orbit, 258
mitigation zone, 39
mobile phone, 249
model, 2, 14, 26, 33–36, 48–49, 51, 115, 125, 204, 214, 218, 220
modelling assumption, 51

modelling crime, 77

momentum, 122

Moon, 107, 132–133, 249, 251, 265

Moon landings, 249

muesli, 232

multi-scale homogenisation methods, 227

N

Napier, John, 53

NASA, 109, 112, 266

National Grid Company, 147, 149–152, 163, 166, 274

natural logarithm, 53

Navier–Stokes equations, 20, 68, 83, 231

Navier–Stokes partial differential equations, 121

NE coast blackout, 148

Neptune, 19–20, 262, 269

network, 147, 151–152, 158, 238

neural net, 17, 173

neutron stars, 271–272

Newtonian mechanics, 19, 270–272

Newton's Cannonball, 254–256

Newton, Sir Isaac, 13–14, 17–18, 20, 55, 68, 76, 122, 254, 259–262, 264, 270

Newton's laws, 20, 64, 66, 70, 262, 270

Newton's laws of motion, 71

Newton's Second Law, 9

Newton's three-body, 263

Nigeria, 180

Nitrogen, 192–193

noise pollution, 38

nonlinear, 187, 219, 221

nonlinear ordinary differential equation, 10, 52

nonlinear second order ordinary differential equations, 76

Northern Hemisphere, 253

North Pole, 110

Norway, 99

nose curves, 166

National Snow and Ice Data Center (NSIDC), 110

numerical analysis, 126

numerical method, 27, 196

O

Oberth, 254

Ohm's law, 153, 160

Oliphant, 198

Omicron variant, 55

Onnes, Heike, 186

operational research, 206

optimal solution, 179

optimisation, 177

orbit, 258–259

orbital velocity, 255

origami, 239–245, 272–273

Oswald ripening, 191

out-of-phase, 73, 77–78

Oxygen, 118

Oxygen-18 isotope, 117

P

paint powders, 233

paleo-climate, 117

paper, 239–245

parabola, 259–260

parameter, 218

Paris Agreement, 134

parking orbit, 258
partial differential equation, 21, 41, 65,
    126, 183–184, 202, 231, 270
patterns, 4
pedestrian, 30–31, 33–35
pedestrian crossing, 30–31
pendulum, 6–12, 16, 71–72, 75,
    77–78, 217, 219–220
pendulum clock, 7
penetration depth, 201
penguin, 193–197
penguin egg, 194–195
penguin temperature, 196
Penrose, Roger, 271
period-doubling, 93
periodic motion, 73, 93
periodic orbits, 100
persistence forecast, 62
phases, 153
Photoelectric Effect, 270
photonic crystal, 225–227
Pippard, Sir Brian, 80
Pisa, 5
Pisa Cathedral, 13
plastic, 214
Plus Magazine, 174, 244
Pluto, 268
Poincaré, Henri, 86–87, 263
policymaker, 50, 134
politicians, 54
polycrystalline, 227–228
polymer science, 231
polynomial equations, 156
positive feedback, 139
potato, 197, 201, 234
power, 152–153, 160, 167
power cut, 64, 94, 163–167

power distribution, 158
power engineering, 156–157,
    161
power generation, 153
power grid, 163, 165
power loss, 154
power network, 158
power station, 148, 154, 160
power supply, 101
power supply network, 161
PP04 model, 136
pressure, 213, 216, 220, 223
Prince Charles, 106
Principia, 254–255
probability, 17, 238
probability distribution, 124
Ptolemaic theory, 260
pulses, 45
Pythagoras Theorem, 2–3, 42
Python, 77–78, 84

Q
quadratic equation, 162–163,
    165–166, 168–169, 174, 219
quadratic formula, 163–164
quadratic map, 90
quadratic ordinary differential equation,
    94
quantum mechanics, 156
quantum theory, 63, 270
quasi-periodic solution, 93,
    220–222

R
radar, 197, 234–236
radar waves, 235
radiation, 198

radio waves, 197, 225

rainfall, 180

Randall and Boot, 198

random, 3, 75, 79

random coefficient, 238

random matrix, 239

random numbers, 100

reactive power, 160

real power, 160

reflection, 41

refrigerated, 176

renewables, 167

renewable sources, 147

resistance, 161

resistor, 238–239

Richardson, Lewis Fry, 122

Riemann Hypothesis, 239

right whale, 39

ripples, 217

risk, 63

Rochester Cloak, 234

rock, 210–211, 213, 217–225

rocket, 254–255, 257, 259

rock folding, 224–225

rock layer, 217, 220, 222–223

Royal Institution, 65, 71, 146, 244

Runge–Kutta algorithm, 84

S

saddle-node, 95

satellite, 108, 248–250, 253–259, 264, 268

saving the human race, 97

Schrodinger equation, 226

Scripps Institution of Oceanography, 114

sea bed, 41–42

sea depth, 45

sea level rise, 111–112

seasons, 251

second derivate, 9

second-order differential equation, 217

sedimentary rock, 213, 218, 222

Seiche Ltd, 45

Seldon, Hari, 24

sensible heat, 186

ship, 38, 44, 46

ship strikes, 39–40

shop, 49

short wavelength radiation, 129

short-wave Solar radiation, 130

Simon Fraser University, 252–253

simple pendulum, 93

Simplex Algorithm, 180

simplify, 12

simulation, 21–22

simulator, 22

sine wave, 222–223

sing, 40

sinusoidal, 152

SIR equation, 83

SIR model, 49–58

Slartibartfast, 99

slingshot effect, 267

sling shots, 266–270

snowball Earth, 142

snowflake, 4–5, 229

soccer ball, 228

sociology, 24

solar, 168

solar eclipses, 251

solar energy, 169

solar panels, 168

solar power, 148

solar radiation, 107, 130

Solar System, 17, 19, 248, 259–270

sonar, 38, 40

song, 40

sound, 45

sound wave, 41

space, 2, 248–250

Spanish Flu, 59

Special Theory of Relativity, 270

specific impulse, 258

speed of light, 66, 271

SPI-M, 60

spring, 211

Sputnik, 255

square root of 2, 243–244

stable, 94, 164

stable fixed point, 94, 100

standard deviation, 101

starch, 203–204

stars, 249

Star Trek, 233

Star Trek: The Motion Picture, 269

statistical approach, 237

statistical behaviour, 238

statistical measurements, 104

statistical model, 173

statistical techniques, 193

statistical test, 189

statistician, 189

statistics, 17

steel, 231

Stefan–Boltzmann constant, 107, 131

St Mary Redcliffe Church, 79

stiffness, 214

stout, 177, 192

straight line, 45

strange attractor, 85–86

structural mechanics, 182

Student's *t*-test, 190

study group, 26, 194

St Valentine's Day storm, 113

Sun, 17, 105–106, 110, 121, 130–132, 137–138, 146, 248, 250–252, 260–261, 264, 266–267, 270

sundial, 252

Sunzi, 251

super computer, 122–123, 128

surveying explosions, 39

susceptible people, 54–55

susceptible population, 50

symmetry, 4–5, 228

symplectic methods, 263

T

tectonic plate, 213

Telstar, 257

temperature, 109, 113, 115, 117–118, 122, 133, 137–138, 140, 183–186, 195, 199, 201, 204–205

tension, 8–9

Tesla, 146, 152, 154

Thanet, 170

The Hitchhiker's Guide to the Galaxy, 27, 99

thermodynamics, 122

Think Tanks, 26

Thompson, D'Arcy, 181

Thorne, Kip, 271

three blades, 171
three-body problem, 262–263
three laws of motion, 7
tide gauges, 111
tides, 251
time, 248
tipping point, 93–96, 106, 110,
    142–143, 148, 163–167
Titanium Oxide, 236
Tizard, Henry, 198
total power, 160
town planner, 86
traffic, 31
trains, 48–49
transmissibility, 50
transmission, 153
transport network, 105
tree ring, 117
triangles, 42
trisecting an angle, 244
truncated icosahedron, 229
turbulent behaviour, 84
two body problems, 18
two-cycle, 89

U
UK4, 125
uncertainty, 126–127
undergraduate courses, 10, 75
universe, 4–5, 9–10, 13, 18, 20, 27,
    271, 274
universities, 48–49
University of Bath, 38
University of Birmingham, 198
University of Limerick, 192
unstable, 94–95, 164
Uranus, 19

V
vaccination, 59–60
van der Pol oscillator, 81
Vaughan, Dorothy, 265
viscosity, 189
viscous, 213
void, 223–224
volcanic action, 138
volcanoes, 116
voltage, 152–154, 165
Voltage Drop, 163, 165
Von Braun, Werner, 255
von Neumann, John, 192
Vostok, 118
Voyager, 259, 269
Vulcan, 20, 271

W
water vapour, 130
wave, 45, 65, 70, 212, 225
wave equation, 40, 212
wavelength, 41, 66, 197, 200, 227,
    235
weather, 62, 72, 74, 82–83, 100,
    105, 126, 173, 177, 180–181,
    249
weather forecast, 16, 62–63, 124,
    126–127, 274
Weiss, Rainer, 271
whale, 2, 27, 38–39, 42, 44–46, 212,
    250, 273
whale song, 41
widget, 193
Wi-Fi transmitter, 81
wind, 146, 168–173
wind farm, 170–172
wind power, 171

wind turbine, 171–172
Wireless World, 258
wood pulp, 239
World Cup, 151
WXCHARTS, 123

X
X-rays, 66

Y
year, 251
younger Dryas, 119
Youngs Modulus, 214

Z
Zeeman, Christopher, 244
zig-zag, 222–223

www.ingramcontent.com/pod-product-compliance
Lightning Source LLC
Chambersburg PA
CBHW050542190326
41458CB00007B/1878